The Behavioural Physiology
of Animals and Man
VOLUME ONE

The Collected Papers of Erich von Holst
Translated by Robert Martin

The Behavioural Physiology
of Animals and Man

VOLUME ONE

University of Miami Press
Coral Gables, Florida

This collection was first published as
Zur Verhaltensphysiologie bei Tieren und
Menschen: Gesammelte Abhandlungen, *Band I*
by R. Piper & Co. Verlag. Munich in 1969
© R. Piper & Co. Verlag
English translation first published 1973
© Methuen & Co Ltd, 1973
Printed in Great Britain by
Butler & Tanner Ltd
Frome and London

ISBN 0-87024-261-X

Library of Congress Catalog Card No. 73-85162

Contents

The papers included in this volume appeared originally in the following journals and under the German titles shown (listed in order of the Contents, above):

Vom Wesen der Ordnung im Zentralnervensystem. *Naturwiss.*, **25**, 625–31 and 641–7. 1937.

Die relative Koordination als Phänomen und als Methode zentralnervöser Funktionsanalyse. *Erg. Physiol.*, **42**, 228–306. 1939.

Das Reafferenzprinzip. Wechselwirkungen zwischen Zentralnervensystem und Peripherie. *Naturwiss.*, **37**, 464–76. 1950 (mit H. Mittelstaedt).

Zentralnervensystem. Das Muskelspindelsystem der Säuger. *Fortschr. d. Zool.*, **10**, 381–90. 1956.

Die Beteiligung von Konvergenz und Akkomodation an der wahrgenommen Grossenkonstanz. *Naturwiss.*, **15**, 444–5. 1955.

Ist der Einfluss der Akkomodation auf bei gesehene Dinggrösse ein 'reflektorischer' Vorgang? *Naturwiss.*, **15**, 445–6. 1955.

Aktive Leiṣtungen der menschlichen Gesichtswahrnehmung. *Studium Generale*, **10**, 231–43. 1957.

Vom Wirkungsgefüge der Triebe. *Naturwiss.*, **47**, 409–22. 1960 (mit U. von St Paul).

Taktile Täuschungen bei Dickenschätzungen durch Betasten. *Pflüg. Arch.*, **275**, 588–93. 1962 (mit L. Kühme).

Vom Wesen des tierischen Lebens. *Kosmos*, 77–80. 1947.

Menschliche Umwelt und Technik. *Gas- und Wasserfach*, **93**, 1–5. 1952.

Glaube, Macht und physikalisches Weltbild. *Merkur*, **12**, 401–5. 1958.

Probleme der modernen Instinktforschung. *Merkur*, **15**, 913–24. 1961.

Über Freiheit. Vortrag für Radio Bremen, 1961. In: Verband Deutscher Biologen (Hrsg.), 4. *Biol. Jahresheft*, 77–81. Iserlohn 1964.

Foreword

In the three decades during which Erich von Holst was scientifically active (1932 to 1962), we can trace the emergence and proliferation of the biological research field best expressed in terms of the concepts 'comparative behavioural research' and 'behavioural physiology'. Within this framework, Erich von Holst was the pioneering founder of *behavioural physiology*.

Behavioural physiology links descriptive behavioural research and physiology. Its approach is based on the question: 'What internal physiological processes of the organism are responsible for a given behaviour pattern and directly express themselves through this agency?' The characteristic method of behavioural physiology consists in investigating certain behaviour patterns, under systematically varied experimental conditions, by means of measurement – that is, *quantitatively*. In the subsequent theoretical considerations, one attempts to interpret the recorded measurements in a logical and mathematical manner. In this context, investigations involving models can be of particular importance. If the theoretical considerations are successful, they frequently open the way to conclusions about certain fundamental physiological functions.

The scientific works of Erich von Holst include detailed original papers on behavioural physiology (i.e. documentary research reports for specialists in the field); essays concentrated on particularly important themes, written in a more illustrative manner for a broader, scientifically educated readership; and, finally, generally comprehensible lectures and other treatises, including consideration of extra-scientific problems.

The two volumes in this edition contain papers from all three categories. In selecting them, the attempt was made to take from the different fields of research those papers which best fulfil the requirements of providing a maximum range of information and presenting material understandable to the interested reader.

A summarizing survey of Erich von Holst's life work will be given, together with a sketch of his personality, in the form of an obituary in Volume II.

Bernard Hassenstein

Translator's introduction

In any developing branch of science, it is usual that a small number of particularly able and outstanding research workers will provide a significant basis which will stimulate a whole network of later, related studies. This is undeniably true of the relatively young science of ethology (the comparative study of animal behaviour), and it is certainly true that the investigations conducted by Erich von Holst acted as a focal point for many of the developments in this field. Strictly speaking, the main area of von Holst's research was that of *behavioural physiology*, but the results of his research were of vital importance to the theoretical foundation of ethology. Ethologists are typically concerned with the observation of the behaviour of intact animals with a minimum of intervention and under conditions which approach those of the natural habitat as closely as possible. In many cases, von Holst's approach differed from this in that he employed active intervention in the central nervous system (CNS) as a means of extracting information about the rôle of the CNS in controlling the behaviour of the intact animal. However, von Holst shared one major area of interest with field ethologists in that he was particularly interested in the *spontaneous* functions of the CNS, in addition to the more widely studied relationship between stimulus and response. One of von Holst's primary contributions to the field of ethology was the demonstration that the CNS is *not* a passive servomechanism linking stimuli to responses, but is a highly active organ, incorporating *numerous functional systems*, which is continuously producing its own stimuli. The point is clearly made that monistic

x

doctrines (such as the all-embracing 'reflex theory') are inadequate to explain the complex interplay of mechanisms within the central nervous system. This is a vital point to remember, since there are still many workers who would attempt to explain the total behaviour of various animals in terms of an aggregation of conditioned reflexes and other relatively elementary components. The question of the contribution of hereditary factors to the behaviour of any animal species is still subject to heated debate, but the papers contained in this first volume of von Holst's papers demonstrate beyond doubt that it is not merely the morphological structure of the brain and its accessory structures which is inherited, but also an entire range of spontaneously active mechanisms which are indispensable to the performance of species-characteristic patterns of behaviour.

It is obvious from the writings of Konrad Lorenz – one of the pioneers of ethological research – that von Holst's investigations of the spontaneous activity of the central nervous system provided an extremely valuable support for the early theoretical framework of ethology. In 1938, in a paper on the behaviour pattern of egg-rolling in the greylag goose (reporting on experiments conducted with another pioneering ethologist – Niko Tinbergen), Lorenz drew upon von Holst's work as evidence for the existence of spontaneous behaviour mechanisms in the central nervous system: 'However improbable it may at first sight appear that finely coordinated and extremely adaptive behavioural sequences of an animal can occur without interaction with the receptors, just like the movements resulting from spinal sclerosis (i.e. that they are *not* constructed from reflexes), there are several important reasons for assuming this to be the case. E. von Holst has been able to provide entirely con-vincing evidence, from deafferentiation studies of the central nervous systems of greatly different organisms, that *automatic, rhythmic stimulus production processes* occur in the central nervous system, *where the impulses are actually coordinated,* such that the impulse sequence transmitted to the animal's musculature arises in an adaptively com-plete form without interaction with the periphery and its receptors.' Thereafter, Lorenz repeatedly referred to von Holst's work on the central nervous system as a source of support for the ethological con-cept of endogenous nervous activity underlying overt behaviour. With-out this neurophysiological support, ethological interpretation based on pure induction would surely have appeared to be much less secure.

In the same way, von Holst was obviously greatly influenced by the early work conducted by Lorenz, and in the more general papers included in the present volume there are a number of references to observations and interpretation stemming from Lorenz's work. It is therefore hardly surprising that this genuinely *symbiotic* relationship between two of the pioneers of ethological theory should eventually have led to more formal collaboration. On 16 September 1958, the Max-Planck-Institut für Verhaltensphysiologie (Max Planck Institute for Behavioural Physiology) was officially opened in Seewiesen-bei-Starnberg in Bavaria, with Erich von Holst as its first director and Konrad Lorenz as his close collaborator. The name chosen for the institute itself reflects the importance of von Holst's neurophysiological approach to the behaviour of whole animals, and the institute remains as one of the monuments to von Holst's scientific works and to his long-term discussions with Konrad Lorenz.

Erich von Holst was one of those rare human beings whose creative activities merit the term 'genius' without a shadow of a doubt. One is particularly struck by two major features of his work which unerringly evoke admiration. In the first place, he had a unique aptitude for remorseless adherence to scientific method in pioneering studies, and yet managed to combine this with a written style of amazing simplicity and clarity. Whilst pursuing a subject in entirely new directions, and acquiring whatever powerful techniques were necessary for his task as he went along, he nevertheless retained the ability to explain in clear terms – with a minimum of confusing terminology – precisely what he was doing, and what the results were. In the present volume, for example, this remarkable capacity is particularly clear from the studies of fin rhythms in medulla-operated fish and from the investigation of the affects of localized brain-stimulation on the behaviour of the chicken. The latter case is all the more astounding in that von Holst acquired, developed, exploited and analysed the necessary techniques as an entirely new development in the last few years of his life. The second remarkable feature of von Holst's investigations is the *breadth of coverage*. It is, at first sight, almost inconceivable that a single man could master so many different areas of investigation in one lifetime, particularly since each study was marked with a special stamp of originality which unfailingly threw fresh light on the field involved. Quite apart from his flair for scientific research, von Holst also possessed great musical aptitude and managed to find time to play the instrument of his choice – the viola – in formal concerts. Indeed, one paper in this present collection, despite its brevity, clearly shows the continually searching and enquiring character of von Holst's exceptional mind. In the course of constructing string instruments as a spare time occupation, von Holst noted that he had a certain difficulty in judging the thickness of wooden laminae through tactile perception alone, and this led to a series of experiments which plainly illustrate both his approach to experimental study and his capacity for continuous generation of key questions leading to further study. A mind of this kind, which is continually questioning, analysing, and producing novel approaches embodies all the attributes of human genius.

Tragically, Erich von Holst died in 1962 from the accumulated consequences of a heart defect which began to afflict him in his teens. Although, at the age of 54, he had already achieved far more than most other men could have hoped for with full health and a lengthy life, his death represented a crippling blow to the emergent science of ethology. Given another decade of active research, there is so much more that von Holst could have done to help lay the foundations of ethological theory. With his accumulated expertise, his gift for research, and his uninterrupted generation of ideas, it is difficult to imagine any limits to the discoveries that he might have made. In 1962, ethology lost a remarkable member of the original group of pioneers, and one can still to some extent feel the vacuum that was left after his death.

I, myself, did not have the privilege of meeting von Holst during his lifetime, and I have only been able to form an indirect impression of him through conversation with his family, former friends and colleagues and through reading of publications. Nonetheless, I feel a deep sense of loss, and having read the present papers in great detail I bitterly regret the fact that I was never able to meet and talk with von Holst himself. It is, of course, a great consolation that one can still experience von Holst through his writings, and I hope that by translating the present collection I will have made his papers accessible to a wider range of people who will enjoy and benefit from them. The task of translation was to some extent facilitated by the clarity of von Holst's papers, but this advantage was offset by the disadvantage of being faced with the extremely wide range of fields which were covered. Every attempt has been made to maintain a high level of accuracy and to preserve the character of von Holst's written style, and I hope that this collection of papers will conform, in this respect, to the high standard which von Holst himself would have expected.

In conclusion, a number of acknowledgements are called for. First and foremost, it should be emphasized that the paper 'On the functional organization of drives' was translated by Dr John Burchard, and was originally published as a translation in the British *Journal of Animal Behaviour*. Since this translation was carried out by Dr Burchard in direct collaboration with Erich von Holst, it seemed far better to reprint it as it stood. The draft translation of the other papers was vetted by Professor Bernard Hassenstein, and I am grateful for various comments that he made. When returning the draft translation with his comments, Professor Hassenstein also provided me with a copy of another translation of the paper on the Reafference Principle, which had been independently carried out by Professor P. C. Dodwell of Queen's University, Kingston, Ontario. As it turned out, the two translations proved to be in extensive agreement, and there was no need for modification of my own draft. However, it was very comforting to see that there was such agreement, despite my own relative lack of experience in the field of biological cybernetics, and I gratefully acknowledge Professor Dodwell's kindness in sending a copy of his own translation to Dr H. Mittelstaedt in Seewiesen for my information.

R. D. Martin, D.Phil.
University College London
July 1973

Part One
Relative coordination

1 On the nature of order in the central nervous system *1937*

In 1894, the physiologist Friedländer performed the following simple experiment: He bisected an earthworm and rejoined the two halves with a piece of thread. When the front half of the worm began to creep forward, the back half was carried along. Surprisingly, the peristaltic waves typical of the animal's movement passed in orderly fashion along the whole worm as if it had not been cut in half at all. The two halves of the worm moved *in coordinated fashion*. From this, Friedländer concluded that anatomical continuity of the ventral nerve-cord is not unconditionally necessary for conduction of nervous excitation. The *'reflex-chain'*, as J. Loeb has called this process, affects one worm segment after another in much the same way as toppling of the first bottle in a row will topple the other bottles, one after the other.

This concept of the reflex-chain rapidly achieved general acceptance since it seemed to facilitate understanding of numerous cases of nervous coordination. One was able to interpret the 'wave'-movements of a swimming eel and a crawling caterpillar, the sequential leg-movements of a scurrying millipede, and even the running of a horse and dog, as a reflex-chain. It was only necessary to assume that each segment, or leg, elicits through a reflex the movement of the segment, or leg, next in the temporal sequence. Even the rhythmic repetition of the same movement, for example the beating of a bird's wing and of a fish's fin, could be understood in a similar manner by assuming that each muscular contraction provides the reflex stimulus for the following contraction of the same muscle or an antagonist. To distinguish such a process from a reflex-chain it was referred to as a 'chain-reflex'. Such reflex performances were, and still are, frequently regarded as the basis for nervous coordination of movement.

In the meantime, however, there has been a rapid accumulation of research workers dissatisfied with this theory. People gradually began to recognize the less rigid, more autonomous variable activity of the nervous system. As a reaction against the chain-reflex theory, one began to talk of the 'holistic character', the 'totality' and the 'plasticity' of the nervous system. In fact, at the present time there is apparently an ever-increasing number of adherents to the view that with the central nervous system (CNS) we are concerned with an organ which – either because of its great internal complexity, or because of the intervention of 'supra-physical' factors – is fundamentally inaccessible to exact functional analysis. For example, one neurophysiologist writes of the 'physically and energetically indeterminable functions' of the CNS. He is of the opinion that this 'functional entity' contains 'its own intrinsic principles, which are inaccessible to purely analytically oriented research', and that it 'appears to be impossible to investigate the intra-central processes with methods involving measurement'.

This paper is aimed at providing a brief survey of my own investigations in this domain, which have led me on to a new interpretation of the nature of nervous coordination.

We can begin with the earthworm as the generally recognized stock example of a reflex-chain. If conduction of nervous excitation along the worm's body is really a reflex-chain, then it must fulfil various conditions. The excitation must always affect one segment *after* the other, just as toppling of a row of bottles makes them fall one after the other. However, under certain conditions (e.g. after the effect of ether) this is not the case. Here, all of the segments extend and contract more or less simultaneously. In addition to this, nervous excitation should not be able to pass along a stretch of the ventral nerve cord on either side of which all of the nervous branches have been severed, since the reflex-chain has been interrupted. But, in reality, nervous excitation passes along such stretches of isolated nerve cord with an actual increase in velocity. Finally, in a section of the ventral nerve cord which has been dissected out of the worm and placed in a physiological solution, the rhythmic contraction impulses should no longer occur, since all possibility of a reflex stimulus has been excluded. In actual fact, the rhythm appears uninterruptedly in such an isolated piece of nerve cord. By recording of the electrical discharges of the ganglion cells, the continuance of the rhythm can be demonstrated over a period of hours (unpublished data). Thus, the central rhythm does not at all necessarily require peripheral stimuli; it is *automatic* and not reflex in nature.

If we consider another presumed reflex-chain, the rhythmic sinuous movements of a fish, we come to the same result. For example, if one takes an eel and severs all the peripheral nerves on both sides of a fair-sized middle portion of the dorsal nerve cord, so that the head and tail sections are only connected by nervous transmission along the dorsal nerve cord these two ends of the body still swim with the opposing order which they exhibited in the intact fish. This occurs even after exclusion of any possibility of mutual mechanical influence, just as if the middle muscular section were still functional. A reflex-chain would not be able to achieve this. And if one operates on a tench, severing all of the dorsal nerve roots on both sides whilst leaving the ventral roots intact (such that no further stimuli can be taken up from the trunk, since the centripetal pathways have been destroyed), the fish is still able to swim around – though its movements are considerably weaker. Thus, in this case too, the impulses which pass to the musculature through the ventral roots are automatically produced in the dorsal nerve cord; they do not require reflex elicitation. Over twenty years ago, Graham Brown performed the same experiment of severing the dorsal roots in mammals, with the same result. We shall later come to another, less crude, demonstration of the automatic nature of the locomotor rhythm. The nervous system is not, in fact, like a lazy donkey which must be struck (or, to make the comparison more exact, must bite itself in the tail) every time before it can take a step. Instead, it is rather like a temperamental horse which needs the reins just as much as the whip.

Objections have been made to the reflex-chain theory from a quite different direction, particularly by Bethe. Bethe saw that in many animals coordination of leg-movements is not rigidly, mechanically determined, but can alter considerably following certain interventions. For instance (taking an example from von Buddenbrock), a stick-insect – like the majority of insects – runs by moving the first and third pair of legs in the same sense, whilst the second pair is always in the appropriate, opposing phase. If the second pair of legs is amputated, the first and third pairs no longer move in the same sense as previously (i.e. in synchronous gait), but work in opposition (alternating gait) as in a lizard or a trotting horse. Guided by the hope that it would be more expedient to determine a certain regularity in such changes in animals with many legs, I studied various centipedes (Chilopoda) and discovered the following with *Lithobius*: The animal normally moves its legs in 'wave'-form. Each leg is separated from the next by a phase-lag of about one-seventh of a step, and each 'wave' thus covers about six to seven legs (Fig. 1.1a). If one amputates one pair of legs after the other, the phase-lag between the remaining legs consistently increases until the maximum of half a step is reached with three and two remaining pairs of legs (Fig. 1.1, b–e). Thus, a *Lithobius* with four legs runs with an alternating gait like a lizard or a trotting horse. What is remarkable about this is that, within certain limits, it is immaterial how great an anatomical gap is left between the remaining legs, whether it involves one, two, three, four or five segments. Therefore, one cannot speak of a fixed reflex locomotor relationship between the legs. A *Lithobius* with three pairs of legs consequently runs like an insect. If one repeats with this 'pseudoinsect' (Fig. 1.1d) the experiment described above for the stick-insect, the same result is obtained: after amputation of the middle pair of legs, the remaining legs are moved in alternating gait (Fig. 1.1c). Overall, one can derive the general rule that the smaller the number of legs present, the greater the phase difference between them. Other arthropods, admittedly with numerous exceptions, generally fit this rule. In more general terms, this rule means: *The processes in one ganglion are quantitatively dependent upon the processes in all other active ganglia.*

One can also reach an exactly corresponding concept of a quantitative reciprocal relationship between the individual parts of the nervous system, by following another tack. It has already been known for about a hundred years that in arthropods extirpation of one half of the brain has the result that the operated animals subsequently run in circles enclosing the intact side. Taking a simple interpretation,[1] each half of the brain transmits to the ventral ganglia excitation which provokes inclination to the corresponding side. If both halves of the brain are active in equal strength, their effects are mutually eradicating; but if one is removed, the effect of the other becomes evident.

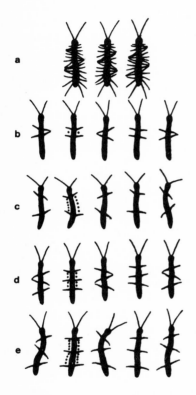

Figure 1.1. Phases from the running movement of the centipede *Lithobius* (from film records). *a*, normal animal. *b* and *c*, after amputation of all pairs of legs except two. *d* and *e*, after amputation of all except three pairs of legs. (The dots in the second diagram in each series indicate the numbers of segments between the legs.)

With centipedes, which regularly run in circles after unilateral brain-removal, one can perform the following experiment. One part (e.g. the last third of the ganglion chain) is put out of action by severing it or simply by cutting off the segments concerned. The result is that the front segments at once twist all the more sharply so that the circles traced in running are reduced in size by almost

[1] In reality, the interpretation is somewhat more complicated, but this is not important in this context.

a half. If the ventral nerve cord is shortened even more from the hind end, the circles become still smaller. This can be exactly measured from tracks left on a powdered substrate. This signifies that in each ganglion of the ventral nerve cord *the strength of influence exerted by the brain increases with decrease in the number of ganglia exposed to this influence.* One might possibly imagine this process in terms of a given 'quantity' of 'brain excitation', which in one case passes to several ganglia and in another only to a few. In the former case, each individual ganglion receives less of this excitation, and in the second case, more. Whether or not this interpretation is apt, there is once again an evident quantitative principle in the reciprocal functional relationships between components of the nervous system.

With the methods so far described – that is, observation of behaviour following amputation of parts of the body and other major surgical interventions – one can, in my opinion, scarcely hope to uncover more than quite general guidelines of neural function, even if these experiments are greatly expanded. It was necessary to find a less disruptive process in order to permit insights into the more intimate processes of nervous coordination.

Before dealing with a process of this kind, we should consider the requirements which a method must fulfil in order to be useful. Difficulties of various kinds combine to bar the way to insight into central nervous processes. The first difficulty is the following: Central nervous function – in this case, the specific patterning of various movements in locomotion – always confront us as a ready-made, harmonic assembly. The pattern is at once present the instant the animal passes from rest to locomotion. It is therefore just as impossible to draw from this ready-made coordination pattern conclusions about the operative factors as it is for the developmental physiologist to draw conclusions from the finished organism about the driving impulses of its development. This difficulty forces us to search for subjects in which the fixed, absolute coordination between the individual locomotor organs is not the only possible relationship, but simply represents an extreme case. (The other extreme would be the lack of any mutual locomotor relationship.) Such subjects could possibly present us with all transitional stages, all degrees of reciprocal influence from the initiation of internal sounding to the strongest manifestation of reciprocal dependence.

On the nature of order in the central nervous system

The second difficulty lies in the multivariate complexity of the processes concerned. One can assume that several different forces, at least, are involved, and a useful method should enable us to extract (from this calculation with too many unknowns) individual quantities for examination.

The third difficulty is based on the well-known fact that in the intact animal no single movement is exactly like any other, so that exact measurements are never entirely successful. This phenomenon is interpreted by some as a sign of supra-physical direction, of free will, and by others as the consequence of continual variation in the external and internal influences upon nervous processes. This persistent irregularity would also have to be excluded in some way.

A process which meets the requirements listed above came to my knowledge at the Naples Zoological Station, to whose director – Professor Reinhard Dohrn – I owe a great debt of gratitude for extensive support and assistance. A fair number of different fish genera (*Labrus, Crenilabrus, Sargus, Serranus* and *Corvina*, among others) are distinguished by the fact that they do not swim along with the usual tail beats but maintain the main body axis immobile and glide smoothly along simply by performing even, rhythmic oscillations of the fins (pectoral, dorsal and anal). From straightforward observation and examination of film-strips, one can see that these individual fin rhythms are performed either in rigid harmony – in *absolute* coordination – or, in some cases only an instant later, completely independent of one another – that is, *without* any coordination. However, what occurs most frequently is that the individual rhythms are indeed exhibited with quite different, independent frequencies, but nevertheless exert an accompanying *quantitative* influence on one another. This behaviour can be contrasted with absolute coordination as a new type of organization, as *relative coordination*.

There is a quite simple method for recording these movements: The fish is anaesthetized and the dorsal nerve cord is severed transversely at a specific point. With one sweep, this operation alters an unpredictable organism to give a *precision apparatus* whose movements are performed quite regularly as long as the external and internal conditions are maintained constant, but which responds to any influence with a quite specific change in activity. After the operation, respiratory movements are arrested;[1] the fish is placed in a

[1] The incision passes through the so-called respiratory centre.

water tank connected with a water inlet, its body is supported, and the individual fins are appropriately attached to light pen-operating levers. When fish of about the size of a human hand are used, it is sufficient if only two or three fin rays are left out of the entire fin. These are connected to the recording levers, and the other rays are cut away (see Fig. 2.9, p. 41, for this experiment). Each individual fin ray has its own pair of small, antagonistic muscles, which move it to and fro and which are not noticeably hindered by the slight resistance of the recording lever, since they are adapted to overcome far greater water resistance. It should be mentioned that the individual fins, as can easily be observed, are not subject to any reciprocal, external mechanical influences. Between twenty minutes and two hours after the operation, one fin after the other begins to oscillate again, and this persists until the death of the fish (after three to five days with good preparations).

As has already been mentioned, the fins can move at different rhythms, and we must now find out what kind of influence is exerted by one rhythm upon another. Two different phenomena emerge, which will be dealt with one after the other.

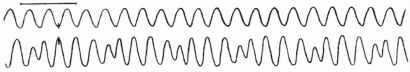

Figure 1.2. Recording of the movement of one pectoral fin (upper curve) and the dorsal fin (lower curve) of *Labrus*. (The horizontal line indicates an interval of two seconds.)

Fig. 1.2 shows a short segment of a curve indicating oscillation of one pectoral fin and the dorsal fin. The upper rhythm is completely regular and independent; the lower exhibits conspicuous periodicity affecting both amplitude and frequency. The external form of these periods varies in a quite specific manner with the frequency relationship between the two rhythms. For the unaided eye, this periodicity is most apparent in cases where the frequency relationship between the two rhythms is exactly 1 : 2, as is illustrated in Fig. 1.3 (a and b). Here, the same picture is continually repeated. (The difference in appearance between Figs 1.3a and b is based on a mutual phase shift in the two participating rhythms, which can be recognized from the coincidence markings.) We shall deal with Fig. 1.3c later on. All such curve segments are immediately reminiscent of periodicity obtained by superimposition of two sinus curves. If such a curve were recorded on a gramophone and played back, the result would actually be the sounding of two notes, e.g. two of an octave in the case of Fig. 1.3. Thus, it would appear *as if* the movement of one fin provides a motor picture of a process of superimposition of two rhythmic excitation processes (of unknown nature) in the spinal cord.

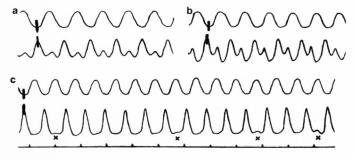

Figure 1.3. Labrus, recording as in Fig. 1.2. All three curve traces are taken from the same fish. (Seconds indicated by the scale beneath *c.*)

In order to evaluate these curves with greater accuracy, I did not in fact make use of a procedure based on the principle of the Fourier series, which is commonly adopted in technology. For the question as to whether superimposition processes of some kind were involved still had to be examined, and nothing of the sort could be presupposed from the outset. Instead, I developed another procedure which leaves the matter of the origin of the periodicity open to all the various possibilities. This procedure provided more exact information about the periodic change in frequency on the one hand, and in the speed of performance of each individual movement on the other, at the same time permitting analysis of the temporal relationships of these variables to one another and to the execution of the independent rhythm (Method of Time and Speed Tables). We shall have to do without a more detailed description here, as it is not altogether necessary for an understanding of what is to follow. It emerges that the periodic departures in the times (frequencies) and speeds, although they are quantitatively variable within wide limits, exhibit

11

extensive qualitative agreement both in their relationships to one another and to the independent rhythm, and are virtually bound to one another. This qualitative agreement is of the kind one would expect in the case of central superimposition of automatisms. One point is particularly worthy of mention: Any increase in frequency of an individual movement is accompanied by a corresponding reduction in motor (contraction) speed, and vice versa. This fact excludes from the outset any attempt to explain the period-formation in one rhythm in terms of a periodically varying general 'excitatory' and 'inhibitory' influence exerted by the other rhythm. At the

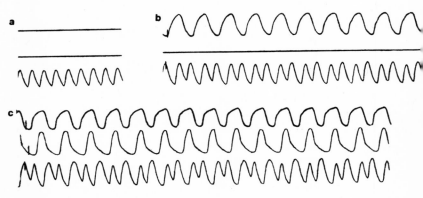

Figure 1.4. Labrus: movement of one pectoral fin (upper trace), the second pectoral fin (middle trace) and the dorsal fin (lower trace). In *a*, the two pectoral fin rhythms are inhibited by pressure on the sides of the body; in *b*, only one pectoral rhythm is inhibited; in *c*, the two pectoral fins oscillate in alternation.

instant when the frequency is 'excited', the contraction process itself evidently undergoes inhibition, and vice versa.

If the superimposition hypothesis is right, it should also fulfil a number of further requirements. As soon as one of the two superimposed rhythms is arrested, the other should at once continue evenly. This is always the case, as is illustrated by the trace in Fig. 1.4. One sees at first (Fig. 1.4a) the two independent pectoral fin rhythms suppressed by means of a pressure stimulus on the sides of the body. Then one pectoral fin rhythm recommences (Fig. 1.4b), and finally both are operating (Fig. 1.4c). In this case, the pectoral fins beat in exact alternation and (as experience has shown) their operation always summates with the dependent dorsal fin rhythm.

With oscillation of both pectoral fins, the effect is in this case roughly 90 per cent greater than with oscillation of just one pectoral fin.

This immediately leads on to another test. If the pectoral rhythms summate in their effects upon the dependent rhythm, as long as the pectoral fins oscillate in alternation, then they would logically have to *subtract* their effects with synchronous oscillation of the pectoral fins and thus cancel one another out by interference. It is, in fact, possible to influence the speed relationships between the individual rhythms in various ways. In the case to be considered (Fig. 1.5), alteration of the oxygen supply has the effect that one pectoral fin beats a little faster than the other, such that it overtakes the other by one beat in the trace shown. At the point where the two pectoral

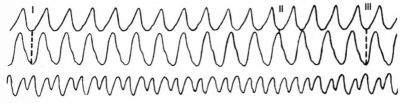

Figure 1.5. Recording of the three rhythms as in Fig. 1.4. At I and III, the pectoral fins beat in alternation; at II they are temporarily synchronous.

fins beat in synchrony (marked by '‖'), the dependent rhythm does in fact temporarily become quite even. This is not because the pectoral rhythms no longer exert any influence but because their individual influences cancel one another.

In the example given in Fig. 1.4, the *independent* rhythm(s) were inhibited (Fig. 1.4a), and the dependent rhythm progressed evenly with its own, inherent frequency. Conversely, one can arrest the *dependent* automatism alone (e.g. by a pressure stimulus on the tail fin). The result is that the fin linked to the dependent automatisms does not remain still, but continues to oscillate with weak amplitude and in full *accord* with the *independent* rhythm. Inhibition of the independent rhythm, by contrast, also causes simultaneous arrest of the dependent rhythm. This experiment further supports the hypothesis of central superimposition and, in my opinion, cannot be explained in any other way.

A further possibility for testing is the following: Fig. 1.6 shows two artificially traced sinus curves with a frequency relationship of

13

1 : 2. One trace increases in amplitude and the other decreases. Superimposition of the two traces initially still gives a large and a small oscillation, then the effect and counter-effect are exactly balanced such that the minor oscillation disappears (indicated by

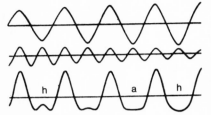

Figure 1.6. Two artificially produced sinus curve tracings, with increasing and decreasing amplitude respectively. Below: superimposition of the two curves (h = zero or rest position).

'a'), and finally the superimposition tracing actually veers in the opposite direction at the point concerned. Thus, if the superimposition hypothesis is correct, under corresponding conditions with increasing strength of the superimposing rhythm there must come a time where one curve peak disappears, and later a point where, at the site where there was movement in one direction, there is actually a movement in the opposite direction. Again, this behaviour could not be explained by 'excitation' and 'inhibition'. Whatever the strength of any inhibition, for example, it could never produce more than arrest and certainly would not produce a directional change in the movement. Let us look once again at the example already given in Fig. 1.3. In Fig. 1.3a, the superimposition is even weaker; large and small peaks alternate. In Fig. 1.3c, there is attainment of exactly the situation where the two automatisms balance one another. Only the characteristic curve form (flattened troughs and pointed peaks) and the minute bumps marked 'x' betray the fact that one automatism is running twice as fast as the other. In Fig. 1.7, one can see along a short stretch of the tracing the transition from almost no influence to the extreme case of directional reversal of the movement. The effect is obtained by gradually cutting off the water; the prepared fish was thus subjected to continuously changing physiological working conditions, which explains the inconstancy of the frequencies and amplitudes. Initially, at the left side of the trace, the central influence of the upper rhythm on the lower one is quite limited. It then gradually increases, and finally, at the points marked 'x' (analogous to the situation on the right of Fig. 1.6), there is an extensive downward excursion replacing the small

14

upward bump. The hypothesis therefore proves to be watertight with respect to this experimental test as well.

In the examples so far considered there was always one independent rhythm and a second rhythm influenced by the former. However, it also occurs that the two rhythms have a reciprocal influence

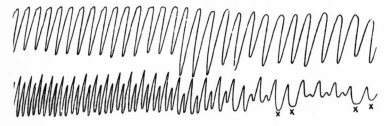

Figure 1.7. Recording of the pectoral fin rhythm (upper curve) and the dorsal fin rhythm (lower curve). The frequency relationship between the two automatisms is 1:2. Water supply to the prepared fish is gradually cut off. See text for further details.

and one can also find a situation where two or three rhythms of different frequency exert an influence on a fourth. In all of these more complex cases, analysis has so far still consistently led to the theoretically expected result. The concept of central superimposition of the different rhythms can therefore be regarded at the present time as a well-supported theory. Thus we can talk, for *illustration*, in terms of *automatic neural 'two-', 'three-'* or *'four-tone' systems*, according to the number of rhythms involved; multi-tone systems which are reflected in the action of the peripheral musculature.

The fact that purely automatic processes are involved in this is indicated, for example, by the observation that one can sever all the nerves leading to a fin exhibiting rhythm-superimposition without abolishing the superimposition effect. The fin concerned is itself arrested, but the continuance of the automatic process underlying its movement remains visible through its superimposition on another automatism linked to a fin which projects *both* automatic rhythms into the external world.

Let us leave the fish for a moment and make a brief excursion, passing over all the other vertebrates to consider *man*. Any human being can without difficulty move both arms either synchronously or in alternation. It is considerably more difficult – for reasons which

15

will be discussed later – to simultaneously beat out different rhythms with the two arms. However, it is usually quite easy to succeed at least in beating twice as fast with one arm as with the other. I therefore arranged for a number of people to perform this type of movement with their eyes closed, registering their activity with appropriate recording-levers. The experimental subjects were not told about the purpose of the experiment; they were simply requested to move both arms as evenly as possible. A perfectly typical example is shown in Fig. 1.8. One of the arms indicated in Fig. 1.8a exhibits

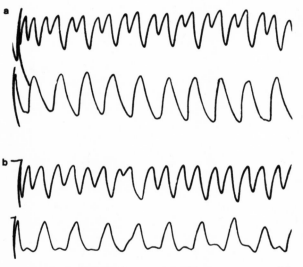

Figure 1.8. Recording of voluntary up-and-down movement of the perpendicularly-held lower arms with a frequency relationship of 2:1. The experimental subject keeps the eyes closed and attempts to move each arm as evenly as possible (above: right arm; below: left arm. Both traces from the same subject).

periodicity of exactly the same kind as that which we have seen with fin rhythms (Fig. 1.3). In Fig. 1.8b, the slight bumps in the lower trace betray the fact that this rhythm was also influenced by the upper one. It is interesting to note that reciprocal influence between the two movements appears, as a rule, to occur subconsciously. The student who produced this particular tracing was afterwards very surprised at its appearance, since he had imagined that he had moved his arms with a continuous, steady amplitude. Of course, we cannot jump to conclusions from these and many other external signs of correspondence, assuming that superimposition is also involved here. This question is still under examination. Nevertheless, they do indicate that these things are not phenomena from some – possibly 'aberrant' – fish nervous system, but processes of general central significance. We shall now return to the dorsal nerve cord of the fish for the simple reason that it is technically much more suitable for investigation.

The resultant concept of automatic, gradually increasing and decreasing excitation processes of some kind, as the cause of rhythmic muscle movements, presents difficulties as soon as one attempts to fit this with what we know of the activity of motor ganglion cells through measurements of action potentials from nerves and muscles. From such measurements, it is safely assumed that the active motor ganglion cell continuously transmits bursts of impulses in rapid succession along the nerve. These salvos of excitation exhibit varying frequencies, and each ganglion cell (as a general rule) possesses its own rhythm of discharge. It is consequently difficult to imagine how the activity of the motor neurons could be organized such that an evenly rising and falling excitation process results in the centre, and how – in any given case – several such processes can be summated. There is, admittedly, a way out of this dilemma, which would be to locate the automatic production of rhythms in *different* central elements and to regard the motor cells as intermediaries between these elements and the peripheral musculature. We shall now have to examine this possibility in more detail.

Even in the course of analysis of relative coordination, phenomena repeatedly emerged which could only be understood through the assumption that the formation and superimposition of the automatic rhythms do not take place in the motor cells which transmit the muscle impulses, but in other regions proximal to these cells. In essence, the following considerations are involved: when one automatism is superimposed upon another, the reflection of this process (the movement trace of the fin concerned) must indicate any increase or decrease in strength of the participating automatisms. However, it emerges that the peripheral activity (the extent of fin oscillation) can increase or decrease within wide limits in response

17

to certain stimuli, without any apparent corresponding alteration of the relevant, underlying automatic excitatory process. An example will illustrate this. In Fig. 1.9, a mild jab with a needle has evoked performance of one double and one quadruple beat of the fin associated with the independent rhythm.

In this case, one would expect that the intensity relationship between the two automatisms should be considerably modified to favour the independent pectoral fin rhythm. In actual fact, however, the steady maintenance of the dependent rhythm indicates that this is not the case. This could be understood on the assumption that

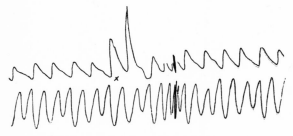

Figure 1.9. *Sargus*: pectoral fin rhythm above, tail fin rhythm below. The frequency relationship is 2:3; the dependent rhythm continuously exhibits triplet periodicity. At *x*, a needle jab evokes two more powerful beats with the pectoral fin. Shortly after this, the recording surface was temporarily arrested (hence the apparent irregularities at that point).

the stimulus magnifying the muscular action has not actually penetrated to the automatism, but has bypassed it to directly affect the motor ganglion cells, and thus the musculature. The converse effect – muffling of the automatic process – can also be evoked. In other words, the intensity of muscular activity apparently proves to be independent of the intensity of the automatism, within wide limits.

Whilst searching for a possibility to settle this dangling question in a more decisive manner, I uncovered the following phenomenon: The tail fin of the fish investigated does not oscillate from side to side as an entire surface; the upper and lower halves of the fin move in opposite directions – when one moves to the right, the other moves to the left. This behaviour can be exploited to provide an answer to our question, provided that one can successfully determine whether these oppositely oscillating fin-halves are underlain by two similarly opposing automatic processes in the spinal cord (case 1), or whether there is only one single tail-fin automatism (case 2), in response to which the musculatures of the two fin halves exhibit opposing (reciprocal) effects through the agency of some transmission mechanism. A decision is obtainable through the superimposition phenomenon explained with the schema in Fig. 1.10.

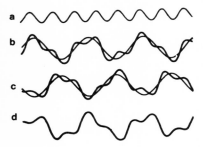

Figure 1.10. Schema for explanation of Fig. 1.11. (See text for details.)

The finely developed sinus curves (b and c) would, in case 1, indicate the opposing passage of the two tail-fin automatisms. The pectoral fin automatism (a) would be superimposed upon them. From the heavily outlined curves of b and c, one can see that the fluctuations evoked by rhythm a always occur in the same direction with both rhythms, b and c. Therefore, for case 1 one would expect corresponding, accordant fluctuations in the oscillatory movements of the two halves of the tail. In case 2, there is only one tail-fin automatism (indicated by the faint trace in c) and, correspondingly, only one superimposition process (heavy trace in c). One half of tail-fin activity would have to correspond to this heavy curve c, whilst the other half of the tail would have to perform, at any given instant, an opposing, mirror-image action – as indicated by trace d. The behaviour actually observed is shown in Fig. 1.11: the traces are perfect mirror images, even covering quite minor irregularities! Thus – assuming that the superimposition theory is correct – it is proven that one-and-the-same tail-fin automatism brings about operation of the two halves of the tail fin. If this is the case, one can only understand the opposing oscillation of the two halves by assuming that between the automatic 'stimulus' on the one hand and the musculature on the other there are further, intercalary neural elements which fall into two groups, which exhibit contrasting (reciprocal) response to the same automatic 'stimulus' and give rise to the corre-

19

sponding behaviour of the two fin halves. Thus, the 'automatic' elements cannot be the ultimate motor elements from which the muscle impulses emerge. In other words: *The automatism is extraneous to the*

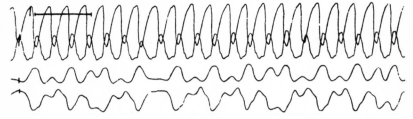

Figure 1.11. Recordings of the two pectoral fins, oscillating in alternation (upper curve), and the upper and lower halves of the tail fin (lower curve). (A central section of the tail-fin rays has been cut away.) Coincidence markings are given on the left-hand side.

'final common pathway' along which all excitation must be passed in order to reach the musculature.

This dualism of the automatic and motor functions of the CNS represents *division of labour*. The automatic elements – we can briefly refer to them as automatic cells – determine the rhythm, the frequency of the muscular activity to be performed. The motor cells are the transmitters of the command, but they (more exactly, the number of impulses which they transmit) determine the intensity (amplitude) of the movement. The motor cells are not only dependent upon the automatic cells; they serve several masters. We have already seen that they can respond directly to peripheral stimuli, which do not even penetrate to the automatic cells (Fig. 1.9), and in the intact animal they are doubtless also accessible to impulses from the brain. Conversely, there are naturally additional effects which only influence the automatic cells, and which therefore only influence the tempo of the rhythmic movements. We shall consider such influences in more detail later on.

For the moment, we shall leave the description of rhythmic phenomena for a while, and turn to an experiment which initially appears to be far removed from the central theme, but whose interpretation leads back to the concept of superimposition.

It is a well-known fact that in all vertebrates normal postural orientation is guaranteed through the activity of the labyrinth system. All authors agree that this applies equally to fish. After destruction of the labyrinth, a fish is no longer able to maintain its equilibrium. However, I discovered that this statement is only conditionally valid. If a fish lacking its labyrinth is placed in a tank illuminated only from above by one light source, the previously disoriented fish once again swims around in normal fashion, as if it were intact. If the light is then projected into the tank *from one side*, such fish all immediately lie on their sides; and if the light source is placed *below* the glass tank, they all swim upside down, with their bellies pointing upwards. In other words: for a fish lacking a labyrinth, 'above' and 'below' is exclusively determined by the direction of incident light – the dorsal surface is always turned towards the light (von Buddenbrock's 'dorsal light reflex'). Thus, for a labyrinth-less fish, the eye provides a replacement static sense, and one would naturally like to know whether such optical equilibrating orientation is operative prior to the suppression of the labyrinth. This question can be answered quite easily. One only needs to illuminate a normal fish from one side. The result is that the fish does, in fact, incline to that side; but the inclination is not complete – it only occurs to a certain degree. The intensity of this lateral tendency can easily be determined. Measurements show that the inclined position exhibited with lateral illumination is maintained just as persistently as the upright position of a fish oriented exclusively by static means. The *degree* of inclination itself is quite exactly dependent upon the light intensity. The greater the intensity of the laterally incident light, the greater the degree of inclination of the fish. Fig. 1.12 shows, for four fish species, the inclination of the body (measured in degrees) in relation to the proximity of a lateral light source. As one can see, each fish species exhibits a characteristic inclination curve – a phenomenon which is extremely interesting from a comparative physiological and general biological point of view, though space does not permit fuller consideration here. The curves clearly show that the normal posture of the fish is a labile feature which is determined at any given time as a *central resultant* of the interaction of these two (apparently extremely different) forces – static and optical 'excitation'.

For an understanding of this central process, the following fact is important: If the fish is kept for a number of hours in darkness and is subsequently illuminated from one side, it at first swims

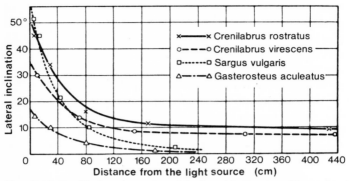

Figure 1.12. Graphical representation of the lateral inclination of free-swimming fish with lateral illumination from a 40-watt-bulb. The degree of inclination is dependent upon the proximity of the light source and there are marked quantitative differences between the four fish species investigated.

exactly upright, with no apparent dependency upon the influence of light. It is a matter of minutes before the fish increasingly inclines towards the light, and the curve of this increase in inclination of the body exhibits a quite characteristic form. The optical equilibration component at first increases rapidly, and the fish inclines more and more to one side. Further increase in inclination then occurs at an ever-decreasing rate, and the ultimate condition of equilibrium between the two forces is only achieved after a number of hours – even days. On the other hand, with fish kept under illuminated conditions and then transferred to darkness, the changeover to exclusively static postural orientation is much more rapid and exhibits a different curve form. A period of 30–50 minutes in darkness suffices to prevent an immediate postural influence of light when the fish is re-exposed to illumination.

This whole phenomenon requires more detailed analysis in all possible directions. For example, it would be informative to study the influence of temperature upon this gradual conversion process and upon the behavioural resultant in general, or to investigate the effect of various substances which affect central activity, and so on. Whatever such experiments may show, it is reasonable to assume that there is, as with the previously described rhythmic processes, superimposition of two 'excitation processes' or 'substances' of an unknown nature. This provides the simplest explanation for the resultant character of the observed behaviour. There is no fundamen-

On the nature of order in the central nervous system

tal distinction in the fact that we are concerned here with persistent conditions, whilst previously rhythmic processes were involved. The fin automatisms themselves are, in fact, not always uninterruptedly active in rhythmic fashion; situations can arise where there are pauses of several seconds between one beat of a fin and another, during which the superimposition phenomenon is approximately maintained. Superimposition in movement is then translated into superimposition in posture.

So far, we have only become acquainted with one of the two processes which play a part in the coordination of the objects of our investigation. Superimposition permits *quantitative* reciprocal adjustment of activities, regardless of whether these exhibit concordant or distinctive rhythms. What it cannot achieve is enforcement of the tempo of one rhythm upon another. Exactly as with a duet, each of the two automatisms remains quite uninfluenced by the other. However, the first striking property of coordinated movements is the accordance in tempo; somehow they are linked to one another. We shall now take a close look at the force behind this property.

When a father goes out for a walk with his six- to eight-year-old son, one can often observe the following: The boy would like to keep pace with his father, but this does not work for long. After a number of coincident steps, he gradually loses the tempo, and in order to fall into step once more he makes one or two rapid additional steps. By doing so, he once again falls in with his father's pace, and the game is all ready to start again. Exactly the same

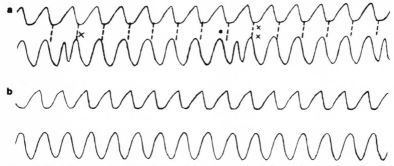

Figure 1.13. Labrus: pectoral fin rhythm (upper curve) and dorsal fin rhythm (lower curve). *a,* relative coordination; *b,* absolute coordination (see text for details).

23

effect – translated into exact graphical form – is shown in Fig. 1.13a. One rhythm is even and independent, while the other exhibits a certain periodicity which is almost exclusively evident in the frequency and less obvious in amplitude. If we take the point marked 'x' as the starting-point, with the aid of the reference lines we can follow how the dependent rhythm gradually becomes displaced with respect to the independent rhythm, until the phase relationship is reversed (•). There is then a much more rapid 'additional beat', which achieves a return to the initial situation (x̄). Thereafter, the periodicity is repeated. A short while afterwards (Fig. 1.13b), the dependent rhythm has become completely aligned with the independent rhythm; the additional beats are lacking, and both rhythms exhibit the same tempo. *Relative* coordination has given way to *absolute* coordination. Fig. 1.14 shows even more clearly that the influence concerned exclusively, or almost exclusively, affects the fre-

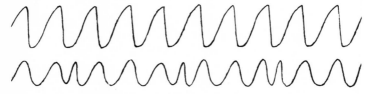

Figure 1.14. A further trace from the same prepared fish as that providing Fig. 1.13.

quency. In this tracing, the frequency varies by a factor of two, whilst the amplitude of the movement remains almost constant. This alone demonstrates an externally obvious, fundamental difference between this phenomenon and that of superimposition, with which variation in amplitude is the first feature to strike the eye.

A further example in which the dependent rhythm is slower than the independent one is shown by Fig. 1.15. In this case, the dependent rhythm does not exhibit occasional additional beats in order to keep pace with the other rhythm, but exhibits the opposite effect of periodic *omission* of a beat. In 1.15c, accordance is finally achieved, and relative coordination gives way to absolute.

The fact that this phenomenon must involve something fundamentally different from the superimposition phenomenon is most clearly indicated by evaluation of such trace recordings using the method of 'time and speed tables' mentioned above. To mention just one important point, any increase in frequency in these cases is accompanied by corresponding increase in the speed of movement, and any decrease in frequency is associated with reduction in speed, whilst with superimposition exactly the opposite occurs. Detailed analysis of this coordination process has so far yielded roughly the following information: Both automatisms – dependent and independent – are involved in its production. The independent automatism operates by exerting a rhythmically rising and falling influence on the frequency of the dependent one; whilst the latter exhibits variations in accessibility to this influence, again with rhythmic fluctuation in degree. The influence itself varies according to the momentary phase relationship between the two rhythms. When close to

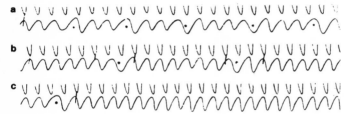

Figure 1.15. Serranus: pectoral fin rhythms (above; truncated to save space) and tail fin rhythm (below). The pectoral fin rhythm progresses smoothly; the tail fin rhythm accords with it periodically and exhibits occasional marked retardation (●). In *c*, from the coincidence mark onwards, the tail fin rhythm is finally in full accord (absolute coordination).

a given reciprocal position the effect is one of attraction and attachment, whilst approach to the opposite position is accompanied by a repellent effect. Because of this polar opposition in functioning, I have proposed the term *'magnet effect'* for this peculiar process – though this, of course, implies no accompanying supposition about its physical nature, which is as yet indeterminate. Briefly stated, therefore, *the magnet effect is the 'endeavour' of one automatism to impose its tempo and a quite specific reciprocal phase relationship upon another.*

If we now enquire as to the reciprocal phase relationship which is imposed by one rhythm upon another, the question can be answered through the intermediary of the superimposition phenomenon. So far, we have only seen examples in which superimposition or the magnet effect occurred *alone*. However, such cases are far from typical. This behaviour frequently occurs temporarily, or only under certain working conditions. As a rule, superimposition and the mag-

25

net effect operate together; but both phenomena can exhibit independent variations in intensity. When the two forces operate together, one can determine a quite specific, invariable relationship between them. The order into which the magnet effect seeks to align the two rhythms (as is finally achieved from 'x' onwards in the trace of Fig. 1.16) always exhibits the same phase relationship as that in which one rhythm augments the other in superimposition. Conversely, the phase relationship which results in *repellence* is the same as that in which superimposition produces weakening of the dependent rhythm (the places marked ● in Fig. 1.16). This, presumably, can only mean one thing. The magnet effect produces reciprocal stabilization with a phase relationship in which both central automatic processes are similarly oriented and *operate in the same sense* – since it is only then that *addition* of the effects can occur with superimposition. Thus, the magnet effect only provides the superimposition

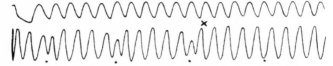

Figure 1.16. Labrus: pectoral and dorsal fin rhythms. Absolute coordination exists from 'x' onwards.

with a biological function when it continually stabilizes the two rhythms with a reciprocal relationship where one augments the action of the other.

One peculiarity of the magnet effect must be remarked upon. The manner in which 'attraction' or 'repellence' is exerted cannot be simply regarded as 'excitation' or 'inhibition'. For example, if the dependent rhythm is, at a given moment, temporally in advance of the phase relationship aimed at by the magnet effect, the latter has a slowing ('inhibitory') effect; whilst if it is falling behind, the effect is one of acceleration ('excitatory'). It is, so to speak, all the same to the magnet effect whether it operates in the form of excitation or inhibition; both influences are apparently unified and the nature of the magnet effect is not encompassed by such a distinction. The magnet effect represents the form in which one automatism fights for control of another.

This struggle can remain undecided for some time, but with augmentation of the magnet effect it must, after a given instant, lead to defeat of the dependent rhythm, to surrender of its independence. Such augmentation of the magnet effect can also be brought about by certain peripheral stimuli. The question now arises as to whether, following the definitive victory of the magnet effect (i.e. after emergence of absolute coordination), the dependent rhythm gradually ceases to strive for independent activity with the passage of time, or latently persists with the struggle. An answer is provided by the experiment illustrated in Fig. 1.17. Absolute coordination exists between the two rhythms; the lower, dependent rhythm had been forced to fall in with the upper rhythm more than one hour previously. Prior to that, the dependent rhythm possessed a frequency which was one-third faster. In the trace (from x to x), the independent pectoral fin rhythm has been stilled by pressure on the anterior sides of the body, and one can see how, during this period, the temporarily released dependent rhythm actually does continue with its

Figure 1.17. Labrus: pectoral and dorsal fin rhythms. Between x and x, the pectoral fin rhythm is inhibited. (The corresponding segment of the dorsal fin rhythm is marked with a horizontal bar.)

tempo increased by one-third. This apparently absurd behaviour, whereby the same stimulus inhibits one action and simultaneously accelerates another, can easily be explained through the disappearance of the magnet effect at this point. From this experiment, which is at any time reproducible, we can only assume that the struggle between the two automatisms persists even after achievement of absolute coordination, though this is unsuccessful for the dependent rhythm and therefore invisible to us. Perhaps for the first time we can catch a glimpse of the tensions which are continuously operating in the CNS without emerging visibly to the exterior; tensions like those which many people must have discovered, through introspection, in their own psychological experience.

In this case, only two automatisms are in conflict with one another. The situation immediately becomes more complex, however, when three or more automatisms are involved and each attempts to maintain its own tempo, particularly when the superimposition phenomenon is added to this. This produces an unimaginable wealth of possible combinations, and corresponding to these there are just as many possibilities of variation in periodicity. The two examples given in Fig. 1.18 are simply intended to demonstrate that such periodicity always permits recognition of a rigid governing principle, which enables us to determine the nature and intensity of the various operative forces. There is, as it were, a parliament continuously sitting in the spinal cord, where each power makes its contribution of force: a parliament which continually reaches

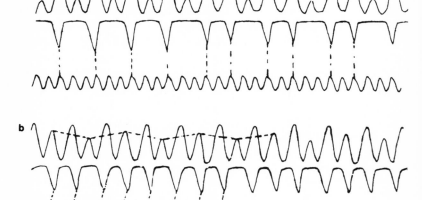

Figure 1.18. Sargus: periodicity produced through the combined operation of the magnet effect and superimposition. (The dotted lines are included to facilitate examination of the periodicity.) *a*, transition from one order to another; *b*, a third combination. The traces represent selected examples from a large number of different types of periodicity exhibited by the same prepared fish.

compromises and which is only really distinct from its human counterpart in that its decisions are never postponed! That ability is a special gift of the brain.

I am inclined to believe that the magnet effect represents an important instrument of central order. In fish, it emerges not only in the interplay of locomotor rhythms but also in the relationship between such rhythms and the respiratory rhythm. With terrestrial vertebrates, this effect is almost certainly the basis for the fixed, absolute locomotor order which characterizes these animals. When we attempt to move our two arms with different rhythms, we are trying

28

to overcome this magnet effect. In fact, some people can win the battle after some practice, at least as far as the arms are concerned. If we lie on our backs and attempt to do the same with our legs raised, we are not usually able to move them with different rhythms. Apparently, the magnet effect of the leg rhythms is stronger than that of the arm rhythms in human beings.

However, the magnet effect is not only the principle effecting order between the different individual automatisms. It seems to operate *within* each automatism as well, i.e. between the individual automatic cells united into a group. This is most obvious when the harmony within an automatism is temporarily removed, e.g. under the effect of anaesthetization or carbon dioxide administration. The entire fin then exhibits an irregular, fluttering movement. Yet if one records the movement of two adjacent, isolated fin rays, they are frequently seen to oscillate quite regularly; but each ray has its own, distinctive tempo. Thus, a group of otherwise cooperating automatic cells has in this case been dissolved to give smaller, independently operating units. Under suitable conditions, superimposition and the magnet effect can be demonstrated in the interaction between these lesser rhythms. Thus, within an automatism there is probably operation of the same process as that operating between various, more distinct rhythms. Since there is usually quite rigid coherence between the elements forming an automatism, the automatism as a whole generally follows the All-or-None Law, i.e. it either discharges completely or not at all. This phenomenon, which was not to be expected in the CNS, is reminiscent of the same kind of regularity in vertebrate hearts; in fact, it leads on to a whole series of additional features which one had previously identified only in the heart, and which provide a kind of bridge between the coordination types found in the heart (so far always treated as a separate category) and those found in the spinal cord. We shall not deal any further with this correspondence at this point, but finish off our excursion through the colourful and varied field of relative coordination with a brief forecast.

Relative coordination is a kind of neural cooperation which seems to be very important since it to some extent renders visible in peripheral muscle action the operative forces, which would necessarily remain invisible given absolute coordination. However, relative coordination is not only of interest in its own right; because of the great precision with which it works we can set questions using the prepared animal and obtain a clear-cut answer. Let us take just one example, concerning the problem of *gaits*.

It is a well-known fact that horses and dogs exhibit various gaits: walking, trotting and galloping. Such gaits also occur in fish. For example, the pectoral fins as a rule beat in alternation, but given certain central conditions they can abruptly 'switch' and beat synchronously. The converse also occurs. We can now ask: On what mechanism is this switch in the internal order, accompanying change from one gait to another, based? There are two possibilities: Either (1) the relationships between the two automatisms remain unchanged, and the response pattern of the automatic cells alters such that they respond to the same automatic process in a manner different to that previously exhibited; or (2) the response of the motor cells remains unaltered, and the reciprocal phase relationship between the automatisms changes. This question must be answerable with the aid of relative coordination. If a third (dorsal fin or tail fin) rhythm is dependent upon the two pectoral fin rhythms, then the influence of the latter would necessarily remain unchanged following alteration of gait in the first case, whereas with the second case an abrupt transition would occur. Fig. 1.19 shows such a case of abrupt, spontaneous change from synchronous to alternating oscillation of the pectoral fins. The third rhythm appears to remain quite uninfluenced until the instant at which the 'switch' occurs. At the moment when the pectoral fins oscillate in alternation, however, the third rhythm is more or less dragged into the new tempo (note the bump in the tracing, marked 'x'), and subsequently it continues with increased intensity of action. Thus, from the point 'x' onwards, the magnet effect and superimposition exert a powerful influence. Since this is *not* the case prior to 'x', one must conclude that the pectoral fin rhythms were up to that point in opposition and had paralysed one another in their effect upon the dependent rhythm. With this, our question is already answered. The change in gait is, in this case, based upon alteration in the reciprocal relationship between the automatisms themselves. In a similar manner, a number of other problems can be tackled with the aid of relative coordination; problems which we were previously unable to examine.

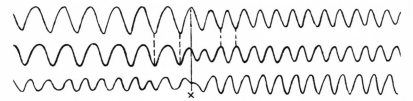

Figure 1.19. Sargus: both pectoral fin rhythms (upper and middle traces) and the tail fin rhythm. Up to 'x', the two pectoral fins beat synchronously (swinging backwards and forwards in unison). At 'x' there is a spontaneous 'switch' to alternating movement (one swings forwards, while the other is swinging backwards). Compare the dotted coincidence lines.

Conclusions

Although the foregoing account presents only a section of the results which have been obtained so far, we can nevertheless draw a number of more general conclusions. *Central coordination is not based upon chain-reflex mechanisms;* it has a quite different nature. Its tools are *processes which only occur within the CNS itself.* The 'reflex' is there in order to adapt this internal process at any given time to varying peripheral conditions, and to alter it in one direction or another. The reflex *is not the basic process itself,* as is so widely believed; it is either an *additional attribute of the central mechanism* or (probably in the majority of cases) a complex interplay of additional mechanisms with the active central forces described above. It is in accordance with the nature of these centrally operative forces that coordination is not fixed and machine-like, but variable, flowing and plastic. But this *plasticity,* contrary to the opinion of many supporters of the 'plasticity theory', is not based on the fact that the CNS is comparable to a nerve net in which 'excitations' spread out, with decrements, on all sides. Instead, it is founded upon extremely refined (one might say *thoroughly engineered) internal organization.* Presumably, morphological structure – the arrangement of the cells and fibre connections – plays a large, but not exclusively decisive, role in the achievement of this order. If it were exclusive, the ordered behaviour at any given moment could not be so very variable. What is important is the variable quantitative interaction of the internal forces (with some of which we have become acquainted). I suspect that these forces are attached to specific cell-types or structures, though at the present time one cannot, of course, draw firm conclusions. Determination of the physico-chemical nature of these processes represents our next task. To accomplish it, the methods described here would have to be combined with chemical and bio-electrical approaches. It remains to be seen how far such a working association will be successful, and what will be discovered as a result.

2 Relative coordination as a phenomenon and as a method of analysis of central nervous functions *1939*

Introduction

This paper represents a report on a group of phenomena involved in rhythmic-motor muscle movements coordinated by the central nervous system (CNS). The phenomena were first observed in various fish species, later in mammals and healthy human beings, and quite recently as effects accompanying certain brain changes found in human pathology. Thus, the common principle is presumably wide-spread. This field of research is in its infancy and can be expanded in all possible directions; nevertheless I am glad to follow prompting to publish a summarized version here. In fact, the results so far available differ conspicuously from what is generally known about nervous coordination of movements in vertebrates. They exhibit a number of parallels to the coordination of quite simple systems (the heart, a medusa) on the one hand and to phenomena of high-level 'psychic' organizational processes on the other; so the object of research would seem to merit a certain interest. The burden of many unanswered questions which is necessarily attached to this essay – and which is not at all concealed – will be compensated if other investigators are successfully induced to occupy themselves with these remarkable problems.

As far as the *object,* and not the *process of evaluation,* is concerned, the investigations do not incorporate any methodological novelties, apart from the apparently unimportant exception of employing organisms which are spontaneously and persistently in action.

33

Usually, the principles of spinal cord function are studied using inactive preparations. This latter procedure is generally preferred for two reasons. Firstly, the popular view is that the reflex action of a previously immobile preparation can be directly related to the effect of the known, carefully gauged stimulus (thus representing a simplification of the process of interpretation); whilst with an object already in action the administered stimulus would encounter a far less transparent central situation, and its effects would be more difficult to trace. Secondly, there is the more theoretical concept that the motor patterns and behavioural activities of intact animals and human beings are composed of reflex elements of this particular kind. Both concepts, as will doubtless clearly emerge in the following account, are evidently inadequate. The CNS does not seem to incorporate under physiological conditions any clearly definable 'resting' condition, even when there is immobility of the motor elements. Even in simple central systems there is continual activity of some kind. As far as the second concept is concerned, the attempt to interpret even quite simple coordinated movements as 'reflex' sequences is ineffectual; instead, we are confronted with certain central forces, whose activity – at least in principle – is independent of any peripheral, afferent influence. The role attributable to the 'reflexes', in view of this fact, will be discussed at the end.

The main task of this paper is the presentation of a summarized account of information concerning the imputed central organizational forces so far derived from the phenomenon of relative coordination and of the conclusions and theoretical considerations which have emerged.

First of all, there is a comparative survey of the important phenomena. This is followed by a description of the general behaviour of the main objects of investigation – fish – and an account of the general and specific principles found to be involved. Analogous phenomena in mammals and human beings are considered next, and, finally, the conceptual picture extracted is contrasted with a number of classical concepts relating to the physiology of the CNS.

A survey of the wide occurrence of the phenomena

The concept of *relative coordination* can be used to cover all central nervous processes, or neurally induced movements, which are composed of simultaneously executed component activities, and which are neither completely independent of one another nor linked in a certain fixed mutual relationship. As far as coordinated locomotor movements are concerned, definitely coupled component actions have long been recognized. In human walking and running, in the different gaits of mammals, in hopping and flying of birds, and (as far as I am aware) in all known forms of movement of lower animals down to the worms, the various locomotor organs operate with a compatible frequency and with specific reciprocal phase relationships.[1] This common type of coordination can be referred to as *absolute*, to distinguish it from relative coordination.

The other extreme, *complete lack of interaction*, is extremely rare.[2] Under experimental conditions, it has been observed with animals possessing many legs (centipedes and millipedes) after amputation of a certain number of legs in the middle. The remaining anterior and posterior extremities subsequently move as two extensively independent groups, with different rhythms (Bethe and Thorner, 1933; von Holst, 1934b, 1935d). Complete lack of interaction also occurs as a normal phenomenon – though infrequently and usually as a fleeting effect – in fish. In human beings, mutual independence of individual rhythmic-motor movements has been observed in cases such as cold-induced shivering (Denny-Brown *et al.*, 1935) and tremors resulting from Parkinson's disease (Jung, 1939). In voluntary movements, it is usually the product of persistent practice (playing of the piano and violin). However, there are considerable individual variations; some experimental subjects succeed immediately and effortlessly in producing rhythmic movements of the two arms with independent rhythms (von Holst, 1938b).

Relative coordination of rhythmic movements was first identified in a number of fish species whose locomotion is represented by oscillation of various fins, with the body usually kept inactive (von Holst, 1935–8). Subsequently, it was discovered in mammals (particularly

[1] An exception under normal conditions is only provided by certain millipedes (*Geophilus*), in which there is rigid coupling between successive legs on one side of the body, with all legs following the tracks of the first, but where the rhythms of the right and left series of legs are mutually independent and adapt to the prevailing substrate conditions (von Holst, 1934a).
[2] Only in lower forms of animal life in which the CNS plays no decisive role in coordination of the parts (e.g. echinoderms) is such lack of interaction evidently frequent.

dogs) as a rare, but quite normal, motor pattern existing alongside the usual absolute organization, and in human beings as an involuntary phenomenon accompanying voluntary movement (von Holst, 1938b). Recently, it has been found in the involuntary, continuous tremors of people suffering from certain brain disorders (Jung, 1939).

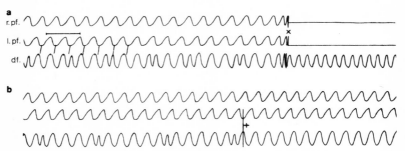

Figure 2.1. Labrus: right pectoral fin (r.pf.), left pectoral fin (l.pf.) and dorsal fin (df.). *a* and *b* are both from the same individual. In *a*, x marks a point where the recording surface was temporarily halted and the pectoral fin rhythms were arrested for some time by momentary pressure on the anterior part of the trunk. In *b*, at the point marked +, there was a spontaneous transition from relative to absolute coordination. (In this and subsequent specimen traces, an upward sweep of the curve indicates a forward movement for the pectoral fin rhythms and a movement to the left for the dorsal and tail fins. The horizontal bar on the left-hand side of the trace indicates a time-interval of two seconds for all curves. Coincidence marks are usually inserted on the left side of the traces. All traces should be read from left to right.)

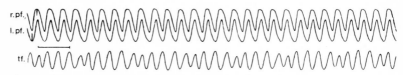

Figure 2.2. Sargus: right and left pectoral fins and the tail fin (tf.).

Figs 2.1–2.8 provide typical examples of relative coordination. Figs 2.1–2.3 are records of medulla-operated fish.

In Fig. 2.1, one can see that two of the fin oscillations take place with a steady rhythm, whilst the third has a periodic form. There is a conspicuous alteration in frequency, which regularly departs by about 35 per cent from the mean value, in either direction. From the point x onwards, where both regular rhythms have been arrested by an inhibitory stimulus, the frequency of the third rhythm remains uniform, approximately preserving the previous mean value. Fig. 2.2 shows another phenomenon. Here, it is predominantly the periodic alteration of amplitude in the lower curve which arouses attention. Fig. 2.3 shows yet another case, where the level of the lower oscil-

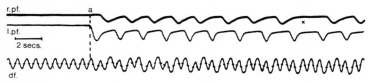

Figure 2.3. Labrus: pectoral fins and dorsal fin. After the point a, there is spontaneous onset of the pectoral fin rhythms. (x marks a spontaneous omission of one oscillation.) The superimposed dotted line on the lower curve is intended to show the level around which the tail fin oscillation varies, following the appearance of rhythmic variations at a.

Figure 2.4. Voluntary movements of both arms from the elbow joint, with the eyes closed. The experimental subject in I (above) was supposed to move the right arm (upper curve) from *a* onwards with gradually increasing frequency, without any influence from the left arm. The unintentional variations in the upper curve are indicated by a dotted line, in the same manner as in Fig. 2.3. In 2.4 II (below), the two arms are supposed to be moved with a rhythmic relationship of 1 : 2. Up to the point *c*, the lower curve is temporarily periodic in form; after *c* the upper curve is periodic (*'Alternans'*, cf. Fig. 2.36 or Fig. 2.48, which show fish rhythms). Between *a* and *b*, a dotted line indicates the form of the curve as seen when the other arm is not moved.

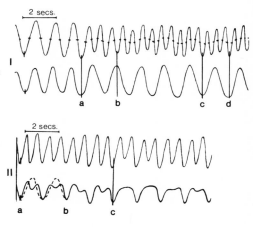

lation, initially uniform, exhibits synchronized periodic upward and downward variations after the onset of two other, much slower rhythms.

A quite analogous example from rhythmic swinging of the arms in a human subject is shown in Fig. 2.4 (I). Variation in swinging of one of the arms in this case appeared involuntarily and unconsciously.

In these examples, only one action has a periodic form; the others are uniform. More often, two or more rhythms exhibit periodicity,

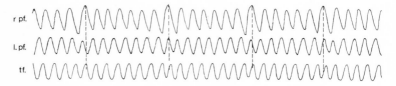

Figure 2.5. Sargus: pectoral fins and tail fin. Relative coordination in which all three rhythms are periodic: the frequency minimum of the upper curve coincides with the frequency maxima of the lower two curves (indicated by the dotted lines).

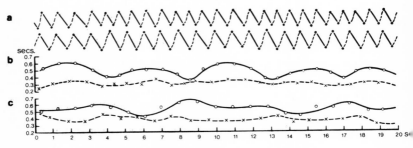

Figure 2.6. Example of relative coordination in mammals, taken from a trotting Alsatian. *a* shows a curve extracted from a film sequence: above – movement of the right foreleg; below – movement of the right hindleg (for every 25 paces of the forelegs, there are 20 with the hindlegs; cf. the film extract in Fig. 2.7). *b* gives a continuous presentation of the duration of each individual movement phase (o–o = thrusting phase; x–x = swinging phase) of the foreleg rhythm. *c* shows the same for the hindleg rhythm. (For the method of continuous time recording, see p. 55 et seq.)

involving certain formal relationships. Fig. 2.5 shows an example from a medulla-operated fish: the upper oscillation becomes temporarily larger and slower at regular intervals, while the two lower oscillations exhibit a synchronous and opposing effect of becoming smaller and faster. Fig. 2.6 provides an analogous example from a free-running dog. The periodic increase and decrease in action times for the movement of one foreleg and one hindleg is shown beneath the general movement curves (see p. 55) for a description of the method). Fig. 2.7, as an illustration, shows a film extract of a dog trotting with relative coordination. In the course of the sequence shown, the forelegs each make $5\frac{1}{2}$ paces, while the slower (but more

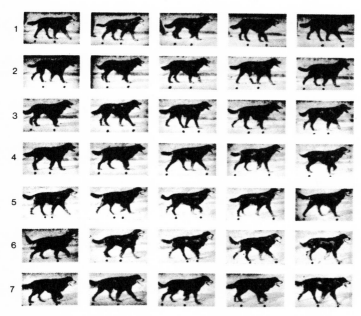

Figure 2.7. Film extract showing a period of relative coordination in a free-trotting Setter. In sequences 1, 2 and 7 there is a synchronous gait; in sequences 3–6 the forelegs overtake the hindlegs by one whole pace. The black spots beneath the two legs on the right side indicate the smooth transition in relative position. (The individual film frames are taken at intervals of half a second.)

extended) hindlegs each make $4\frac{1}{2}$ paces. The relationship starts out in the form of a synchronous gait, and smooth alteration of the frequency alters it first to alternating gait and then back to synchronous gait. In some cases, the same sequence can occur repeatedly for some time.

Finally, Fig. 2.8 shows that one and the same animal may exhibit a wide variety of forms of relative coordination. The nine trace segments are all taken from the same medulla-operated fish; each exhibits a conspicuous, quite regular form of periodicity distinct from all the others (in d, two types of periodicity occur in sequence). The range of possible variations is actually extremely large, as will be shown in more detail later on. Nevertheless, analysis of these phenomena has shown that it is possible to relate them all – along

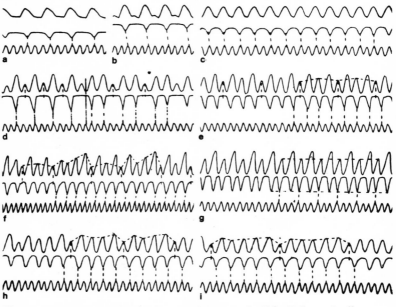

Figure 2.8. Sargus: pectoral fins (1st and 2nd curve) and tail fin (3rd curve): all traces from the same prepared fish. The nine trace segments show different forms of periodicity (in *d*, there is spontaneous transition from one periodic relationship to another). The various dotted lines are included to make the periodicity more easily discernible, and to indicate the relationships between the three rhythms.

with absolute coordination and various well-known special cases (e.g. the phenomenon of 'reciprocal innervation') – to an as yet unidentified factor, whose operation was first clearly exposed through relative coordination effects. This factor can be most clearly derived from curves obtained from fish. We shall therefore turn exclusively to experiments on fish.

Methodology and general properties of medulla-operated fish

According to comparative observations, the phenomena of relative coordination are widely dispersed among teleost fish. However,

40

under natural conditions they are little apparent owing to the general behavioural instability of the fish and rapid changes in internal conditions. When the brain is disconnected, the movements become fixed and regular, without any alteration in their natural properties.

METHODOLOGY

Transection should be carried out in the upper medulla such that the line of incision passes through the posterior, overlapping section of the hindbrain. The incision does *not* have to pass through a *particular* region; nevertheless, there is a general rule that incisions which are too far forward often lead to incomplete regularization of the spontaneous movements, whilst transection which is too far back

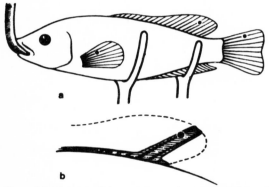

Figure 2.9. Diagram of the experimental set-up. *a, Labrus* with transected medulla. A glass tube with a controlled water current leads into the mouth. The body is fixed, and light recording levers are attached to the fins at the points marked ● Attachment is usually ensured with an intervening lever arm so that fin movement is not hindered anywhere. In *b*, all the dorsal fin rays except two have been cut away. The remaining two continue to beat unhampered.

has the result that the oscillations do not appear for some time (those of the pectoral fins frequently failing to appear altogether) and remain weak. Respiratory arrest regularly appears, and the fish is given artificial respiration (Fig. 2.9a). The body is fixed in an appropriate manner and the fins are attached to light levers such that their action is recorded with a minimum of distortion. In many cases, several rays of a fin are attached to levers; if fish of sufficient

41

size are selected (weight: 200–300 gm; length 20–30 cm), just one fin ray can be used to operate a lever, since the muscles are naturally adapted to overcome a steady resistance (that of the water). Each fin ray is moved by a pair of antagonistic muscles. In the dorsal fin, which is usually the one used for measurements of great exactitude, the muscles of each fin ray are separated, whilst the muscles of the tail fin and pectoral fins are assembled in more compact form. But even with these fins, the fact that neighbouring fin rays can exhibit quite different movements indicates that they are mechanically extensively independent.

The possibility of reciprocal mechanical influences exerted between the fin oscillations by resultant water movements was reliably ruled out by frequent control experiments involving, among other procedures: (1) recording out of the water; (2) recording in water following removal of all the fin surface other than the fin ray operating the lever; (3) artificial oscillation of a fin in a dead, prepared specimen to test whether other fins exhibited accompanying oscillation. The fish used belonged to the genera *Labrus, Crenilabrus, Sargus, Serranus, Corvina, Smaris* and a number of others. The recorded tracings, which provided the basis for this account, amount to a total length of about two miles.

GENERAL BEHAVIOUR: INDEPENDENCE OF THE
INDIVIDUAL RHYTHMS

The activity and response pattern of medulla-operated fish differ considerably from the behaviour generally reported for animals with a spinal cord preparation (including fish), and a brief description is therefore necessary.

Onset of the rhythms; effect of water-exclusion

A fish which has undergone operation and has been provided with artificial respiration at first exhibits a non-excitable period usually lasting about 5–10 minutes. It then passes through a phase in which it behaves similarly to a fish with a spinal cord preparation, i.e. it only exhibits specific 'reflex' fin movements in response to specific stimuli. Fig. 2.10 shows such responses. This indicates that the spinal cord has at least partially regained its sensitivity to excitation, whereas the medulla region has not yet sufficiently recovered.

Sensitivity to excitation then gradually increases such that the same stimulus, a little while later, already evokes a short series of rhythms. Instead of using external stimuli, one can under these con-

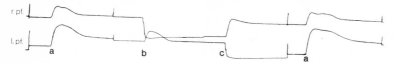

Figure 2.10. Serranus: both pectoral fins. Elicitation of 'motor reflexes' by stimulation of the skin of a fish shortly after transection of the medulla, when excitability is still low. *a,* stimulation of the dorsal fin; *b,* stimulation of the right pectoral fin; *c,* stimulation of the left pectoral fin (all through brief compression with forceps). A rise in the curve indicates a forward movement of the fin.

ditions produce temporary operation of a rhythm in an indirect manner, e.g. by exclusion of the respiratory water (see Fig. 2.11).

Finally (after 20 minutes to several hours), under constant conditions the fin rhythms emerge 'spontaneously' and then continue for several hours or even several days until the fish dies. The initial

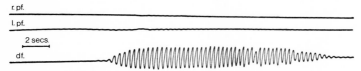

Figure 2.11. Labrus: a preparation which has not yet spontaneously resumed action following the operation. The water supply was excluded from about 70 seconds before the beginning of the specimen trace and continued beyond the end of the trace (right side). The dorsal fin rhythm is temporarily exhibited as the sole response.

emergence is periodic, but later it is generally uninterrupted. As a general rule, each fin begins to operate independently of the others. (In many cases, one or more fins may fail to operate throughout the entire period of the experiment.) Fig. 2.12 exhibits a stage in which one pectoral fin is fully active and the other is operating only weakly, while the dorsal fin is still completely immobile. Accordingly, water exclusion has differing effects on all three rhythms (if we consider just the *amplitude* for the time being): the first rhythm is gradually inhibited; the second is promoted up to a maximum (a) and subsequently inhibited; the third is initially sparked off and then promoted to reach a maximum (b) much later than the second rhythm. Thus, the decisive factor is the condition in which the change (in this case CO_2-enrichment and O_2-deficiency) encounters

43

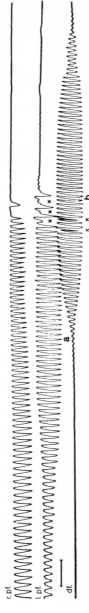

Figure 2.12. Same fish as in Fig. 2.11. Half an hour later, with normal water supply, first the right and then the left pectoral fins have begun to operate. The water supply was excluded 40 seconds before the beginning of the trace, and the three rhythms thus exhibit different responses. The lower rhythm (dorsal fin) behaves as in Fig. 2.11. (There is absolute coordination between the rhythms, with transient relative coordination at the points marked x.)

the ganglion elements in any given case. The reactions of these elements can themselves be utterly opposed to one another under certain conditions. This is a quite general rule in the CNS, which can be drastically demonstrated with medulla-operated fish in many such simultaneous experiments. Consideration of this rule would elucidate many of the apparent contradictions which appear even in the higher behavioural aspects of these animals.

There are certain typical differences between the individual fin rhythms in the manner in which they emerge under constant conditions. The two extreme possibilities are that there is either (a)

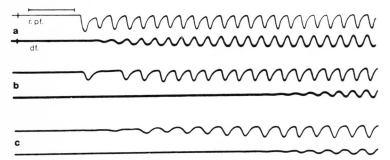

Figure 2.13. Labrus: various forms of spontaneous onset of the right pectoral fin with the same prepared specimen. In each case, arrest prior to the emergence of activity was achieved by temporary closure of the respiratory water supply. Closure lasted for 40 seconds in *a*, 70 seconds in *b*, and 120 seconds in *c*.

gradual increase in amplitude, with the frequency roughly the same from emergence onwards, or (b) constant amplitude from emergence onwards, with the frequency either constant or initially slower and (frequently) not quite uniform. Case (a) is generally the more common; it is always found with the dorsal fin rhythm, often with the tail rhythm, and in some fish (*Corvina, Smaris*) quite frequently with pectoral fin rhythms as well. Case (b) is usually quite purely represented in the pectoral fin rhythms, particularly in *Labrus* (Fig. 2.3), and often occurs in the tail fin rhythm of *Serranus*. In case (b), the rhythm tends to occur either fully or not at all, and this tendency is generally exhibited in relation to various promoting or inhibitory influences. Rhythms of this kind will be referred to in the following account as 'All-or-Nothing' (A.-o.-N.) rhythms. The cause of this difference will be discussed later on, and it need only be pointed out that rhythms of the A.-o.-N. type can also lead on to the other type of action. Fig. 2.13 a–c shows an experimental series in which the pectoral rhythm is induced to exhibit different initial amplitudes following resumption, corresponding to the duration of a previous inhibition. The other fin rhythm (dorsal fin) behaves in agreement in all three cases.

Behaviour with change in temperature

As with the previous examples, extensive mutual independence of the individual rhythms is seen in the pattern of response to temperature change. Fig. 2.14 provides extracts from an example experiment.

The traces a–f show the behaviour of the individual rhythms as the temperature slowly increases from 10°C to 25°C. Only the

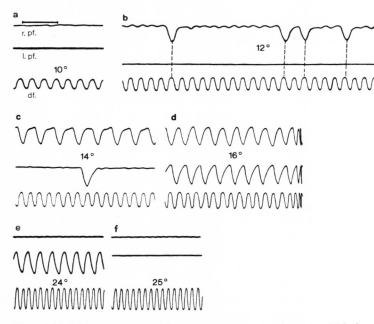

Figure 2.14. Labrus: extract traces from a temperature experiment in which the environmental and respiratory water temperature was raised from 10°C to about 25°C in the course of roughly 60 minutes. The prevailing water temperature (quite close to the fish's body temperature) is given for each extract. The curves show extremely different behaviour for the three rhythms.

dorsal fin rhythm (third curve) behaves in a uniform manner. The pectoral rhythms differ from one another. One emerges at 12°C with sporadic beats and is continuously active from 14°C onwards; its frequency then rises by about a quarter up to 22°C and then disappears at 23°C. The other pectoral fin rhythm first emerges at 14°C with sporadic activity and is continuously operative at 16°C; it persists until 24°C, with an overall frequency increase of about one-sixth. At 25°C it abruptly disappears.

Many temperature experiments show that temperature evidently exerts an influence upon the central rhythms along various pathways, some of which are quite indirect. At least, this is one way of interpreting the facts that different rhythms of one and the same prepared fish can exhibit such differences in dependence upon temperature, and that a change of any given operating condition of the object also takes effect on the form of temperature dependence. In most cases, the fin oscillations exhibit such extensive differences from one another in behaviour that one can scarcely avoid the assumption that each oscillation is underlain by its own central process, which must come to terms with the individual variable factors more or less independently of the others.

Behaviour accompanying strychnine poisoning

The stated thesis of fundamental independence of the rhythms is also evident from experiments on fish treated with strychnine. Fig. 2.15 provides a quite typical example for increasing irregularity of the individual fin oscillations following initial operation of the poison, until there is only sporadic emergence of cramps. Each fin oscillation, uninfluenced by the others, passes through the individual stages of this discoordination process with its own timing. In this instance, it is actually quite common to see extensive variations in movement between different fin rays within one fin (Fig. 2.15d).

Effect of peripheral stimuli

Stimuli act upon a spontaneously active prepared fish in a manner quite different from their action upon a fish which is only activated in a 'reflex' manner. This will be illustrated with only one example at this point, as description of the manifold effect of stimuli will emerge as a side-product from the following account. Instead of the two fin reflexes in Fig. 2.10, which can be evoked with a still inactive medulla-operated fish or a spinally-operated fish (1. forward beat of both pectoral fins with stimulation of the dorsal fin; 2. backward beat of the stimulated fin and forward beat of the opposite fin with compression of one pectoral fin), the first stimulus in this case produces – according to its intensity – partial or complete arrest

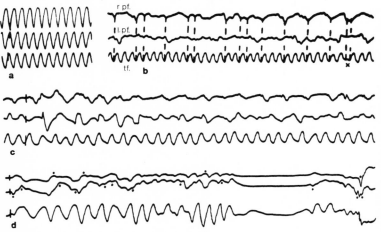

Figure 2.15. Sargus: strychnine experiment (intramuscular injection of 0·9 mg strychnine per 100 gm fish body weight). *a,* 5 minutes after injection – the rhythms are still uniform, but they are accelerated with respect to the normal situation. *b,* 25 minutes later – the tail fin rhythm is still quite uniform; the pectoral fin rhythms are irregular. One can still recognize certain relationships between the upper and lower curves with the aid of the dotted lines. *c,* another 20 minutes later – utter discoordination of the three rhythms. In *d,* 120 minutes afterwards, the two upper curves are taken from two fin rays of the same (right) pectoral fin. The superimposed dots indicate the points where the two fin rays differ in their movements. The lower rhythm is also quite uncoordinated by this stage.

of the pectoral fin rhythms. The two rhythms only respond equally when they are equal in activity; otherwise, it is predominantly or exclusively the momentarily weaker rhythm which is inhibited. With the second stimulus, there is arrest of the stimulated fin, accompanied by continuance of the other fin rhythm with either reduced or (as shown in Fig. 2.16) increased frequency.

With the spontaneously active fish, as can be demonstrated with many additional examples, the same stimulus has a much greater range of possible effects (over and above possible variations in effect through varying stimulus intensity) than is evident in its limited operation on a prepared fish which is not spontaneously active. With the former, the effect can – under certain conditions – be both 'inhibitory' and 'excitatory' (Fig. 2.16), whereas with the latter the effect is more standardized and is often qualitatively different. The basis for this difference only becomes evident with understanding of the internal relationships between the central rhythms.

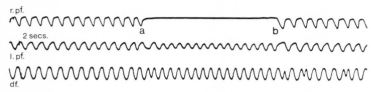

Figure 2.16. Labrus: example of varied effect of a stimulus (pinching of one fin) on the spontaneously active prepared fish in comparison to the limited effect on the inactive fish (cf. Fig. 2.10). Up to *a* there is absolute coordination between the three rhythms. At *a*, there is brief pinching of the right pectoral fin. Afterwards, there is arrest of the stimulated fin and continuance of the other two rhythms with their frequencies increased by about one half (frequencies which these rhythms had also exhibited before the onset of the right pectoral fin rhythm). After spontaneous resumption of the upper rhythm at *b*, there is a return to the previous frequencies by both pectoral fin oscillations. The dorsal fin oscillation passes over to relative coordination.

Analysis of the individual phenomena

COUPLING EFFECTS

A suitable way to begin analysis of relative coordination is to conduct a formal classification of the wide range of periodic phenomena into a number of categories, among which there are two that play an important role: periodicity in which there is a clear relationship or equalization between the various frequencies, and periodicity where the individual rhythms evidently proceed without any mutual influence. We shall first consider the former category, starting with an investigation of cases in which the frequency influence only affects *one* rhythm, whilst the other(s) remain approximately or completely constant in frequency.

General laws

Fig. 2.1 (a and b) provides a typical example. Comparative study of such forms of periodicity using a large number of fish and many hundred metres of curve recordings has led to discovery of the following general rules:

1. Periodicity only emerges when there is at least one other fin rhythm in action, apart from the primary rhythm concerned, and the frequency of this other rhythm must differ from that which exhibits periodicity. Fig. 2.1a shows that the lower rhythm is exhibited quite uniformly, as long as it is either operating alone or is in concordance with the other rhythms (Fig. 2.1b, after +).

2. The rhythm exhibiting periodic variation can be either the faster or the slower of the two. There is no rule that (for example) the more rapid rhythm must always be the uniform one.

3. Any alteration in the reciprocal frequency relationship between the rhythms, whatever its cause – peripheral stimulation or other alteration in ambient conditions (temperature, O_2-supply, narcosis) – is accompanied by a parallel, characteristic alteration in the form of periodicity.

4. Independently of this, the degree of periodicity (= degree of departure from uniformity of appearance) can be varied over a wide range by impinging stimuli and other influences. The most extreme cases of these variations are represented on the one hand by transition to complete uniformity in the frequency inherent in the rhythm, and on the other by increase of periodicity to a maximum, which is abruptly followed by transition to the uniform tempo of the other rhythm, i.e. from relative to absolute coordination (Fig. 2.1b).

5. Conversely, absolute coordination can be converted to relative coordination, of the type considered, by all stimuli or influences which are known to either accelerate or retard the frequency of one of the rhythms involved or which reduce the intensity of activity (measured from amplitude of movement) of the rhythm which remains uniform.

These rules, which apply in the same way to all of the fish genera investigated and to all fin rhythms studied, lead to the conclusion that the resulting periodicity of one rhythm is directly internally correlated with the performance of the other. They thus justify more detailed evaluation of the periodicity in the curves, carried out from this point of view.

General results of evaluating the curves

Whereas uniform rhythmic oscillations are adequately characterized by specification of the frequency and amplitude, the peculiarities of the curves involved here cannot be completely described in this manner. As far as the frequency is concerned, detailed measurement shows that there is not only variation in the duration of complete cycling (complete rhythms measured from one given point to the reappearance of the point selected), but that there is also extensive alteration of the temporal relationship of the oscillations in one direction and in the opposite direction (phases and counter-phases) in many cases (cf. Fig. 2.47d). Thus, when more exact data are required, it is necessary to measure the duration of each individual phase from the point of inflection on one side to that on the other side (*action time*), rather than measuring the frequency. When the action time is fixed, the momentary amplitude of the fin oscillation is determined by the rapidity with which the muscle contracts within a given time. The latter is, in its turn, determined by the number and temporal summation of the motor discharges affecting the muscle fibres. Thus, in the following account, rather than giving measurements of amplitude I shall usually specify in addition to the *action time* (where it is necessary) the *average speed of movement* for each phase (calculated from the angle formed with the horizontal by the connecting line between two points of inflection).

Statistical procedure. A very simple treatment, which is carried out without measurement of action times and speeds, is the following: One relates some chosen, characteristic and recurrent point on the periodic rhythm with a particular selected point on the non-periodic rhythm, e.g. each upper inflection point of one with each lower inflection point of the other, neighbouring curve. From the superimposed indicator lines in Fig. 2.1a (left), it can be seen that the spatial relationships between these two points can vary greatly from case to case. If one now takes a trace of some length (Fig. 2.1a) and measures in each instance how far to the right or left of the nearest reference point of the upper curve the corresponding point of the lower curve is located, it emerges that every possible spatial relationship between the two points does indeed occur, but that a particular reciprocal spatial relationship is clearly most common. If the individual measurements are arranged according to their

51

spatial relationship and entered in a histogram in groups as ordinate points above an abscissa, whose length covers the whole possible range, one arrives at the picture in Fig. 2.17a. The line linking the ordinate points forms a clear-cut, unimodal variation curve, with the notable feature that it does not meet the abscissa at any point. The form of this curve can vary somewhat; one also finds curves which are mildly asymmetrical. However, all curves agree in that *the consistently flattened peak of the curve lies at a specific, well-defined point,* namely at the point where *all* determinant points lie in the case of *absolute*

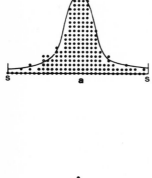

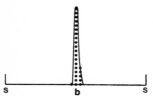

Figure 2.17. a: in a trace of some length, a sample of which is given in Fig. 2.1a, a variation statistics procedure is used to determine the frequency of the various possible spatial relationships between two points (the position of each upper point of inflection of the third curve related to the nearest lower inflection point of the second curve), which is represented as a histogram. The distance s–s represents the entire possible range (i.e. in the middle curve of Fig. 2.1a (left), the distance from one of the superimposed reference lines to the next). The limits of this distance are so selected that the curve maximum in Fig. 2.17a (i.e. the most common phase relationship between the two points) lies in the middle.

b: analogous evaluation of a short trace segment from Fig. 2.1b (right) showing absolute coordination. Here, the relationship between the points of the curve does not vary; it is approximately constant.

coordination. Fig. 2.17b shows the measurement of a short segment of the further passage of the trace (Fig. 2.1b), where relative coordination has passed over to absolute.

The previously variable phase relationships are now extensively constant, and there is only that spatial relationship which was previously extremely common. We can thus define the observed form of relative coordination: relative coordination of this type is a form of organization in which the relationship applying continuously to each component action in absolute coordination only represents *a statistical rule of relationship.*

This statement has the following corollaries: 1. the emergence of periodicity in one rhythm is based upon some kind of relationship to the uniform rhythm (and not, for example, upon rhythmic influences exerted by another frequency differing from this); 2. the factor giving rise to the periodicity, and thus to the statistical rule of co-ordination, is probably the same as that which produces the rigorous phase relationship in absolute coordination. In the former case, however, this factor is possibly subject to some counter-influence which hampers the production of fixed phase relationships.

In a series of cases – namely, in all those in which simple, whole number frequency relationships (e.g. 1 : 2, 2 : 3, 3 : 5, etc.) are established between the uniform and the periodic rhythm – the result of such statistical evaluation departs from that shown in Fig. 2.17a in a characteristic manner. An example is given in Fig. 2.18, which is based on the traces provided in Fig. 2.19 a–c. In Fig. 2.19a, the

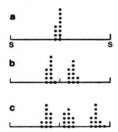

Figure 2.18. Variation statistics evaluation of the three curves partially represented in Fig. 2.19a–c. Here, the lower peak of the first curve has consistently been related to the upper peak of the second. (All other features as in Fig. 2.17.)

upper rhythm regularly omits every second action and the oscillations performed bear a fixed relationship to the uniform rhythm. Evaluation in Fig. 2.18a therefore gives a picture of absolute co-ordination. The relationships in Fig. 2.19b and c are different. In Fig. 2.19b, two oscillations follow one another and every third is omitted. Evaluation in Fig. 2.18b demonstrates that the two oscillations exhibit different phase relationships to the uniform rhythm. The unitary reference points in Fig. 2.18a have divided into two equal groups, shifted somewhat to the right and left of the previous position. In Fig. 2.19c, there are groups of three oscillations occurring with a tempo which is evidently retarded relative to the uniform lower rhythm, thus giving rise to pauses between the 'triplet' periods. Evaluation in Fig. 2.18c accordingly gives three separate groups of reference points. The two groups from Fig. 2.18b have moved closer together and somewhat to the left, and there is an additional third group to the right.

Pictures similar to those in Fig. 2.18 were always obtained from evaluation of cases in which specific, constant frequency relationships were maintained. From the comparison of numerous different cases of such periodicity it is possible to derive the following rules.

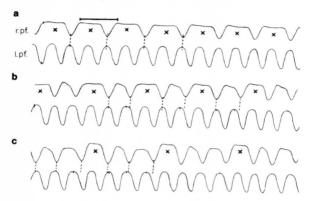

Figure 2.19. Labrus: both pectoral fins. The extracts *a–c* are taken from a long trace of an experiment in which the right pectoral fin was initially arrested by an inhibitory stimulus. After resumption of the rhythms, there were at first only isolated beats. After this, at *a*, there was for some while a beat of the right fin for every second of the left. Later on, at *b*, there were pairs of beats with intervening pauses, and later still, at *c*, triplets. Eventually, there was transition to absolute coordination. (The peaks of the upper curve are directed downwards; pauses are indicated by x. One can see that with decreasing frequency of pauses from *a* to *c* there is also reduction in their duration.)

In the simplest case, with a frequency relationship of 2 : 1 (Fig. 2.19a), the phase relationship of the oscillation performed each time is the same as with normal absolute coordination. In all complicated cases, a varying number of groups of reference points is produced. However, the groups are not uniformly distributed over the entire possible range. Instead, they are most common in the region characteristic for absolute coordination. When one produces a series of different, whole number frequency relationships one after the other with one and the same experimental subject (through exposure to certain stimuli, alteration of the water supply, etc.), and each of the overall groups of points produced is entered as a single point in a common histogram, then one obtains a variation curve like that in Fig. 2.17, with a peak at the point corresponding to an absolute coordination relationship.

From these statements one can conclude that even in cases where whole number frequency relationships of quite varied types emerge, the operative factors are the same as in the other cases. The difference is possibly due to the fact that a stable equilibrium is produced in such instances, whilst under somewhat altered conditions this cannot occur.

Procedure of continuous time and speed registration. A more detailed understanding of the variation in action times and speeds giving rise to the observed form of periodicity permits the following procedure: For each action phase of the periodic oscillation, one measures the duration and the speed of movement (or simply the angle of slant of the corresponding curve), and these values are entered continuously on graph paper beneath the curve. In this way, one can make an overall comparison between the curve tracing itself and the calculated data, which permits easy recognition (for example) of the basis for alterations in the period at any point. Fig. 2.20 provides an example.

If one at first pays attention only to the behaviour of the movement curve of the dorsal fin, one can see that the action times of the phases and counter-phases lie on a common line (lower curve in Fig. 2.20b), which oscillates up and down, initially steeply and then becoming flatter. In the same way, the values for the action speeds for both phases lie on a common line (Fig. 2.20c), which exhibits the same kind of oscillation but represents a mirror-image of the line representing the action times. In both cases, the amplitude of these curves provides a measure of the variation of the two parameters (action times; action speeds). The frequency of these plotted curves is rigidly dependent upon the reciprocal frequency relationship between the original rhythms. In the plotted curves (b, c), an entire oscillation corresponds to one curve period in the fish rhythms, in the course of which one rhythm overtakes another by one whole oscillation. These principles are quite general in application and we can derive from them the following statements: 1. In the observed type of periodicity, an increase in frequency (= reduction in action time) is accompanied by an increase in the speed of movement, and conversely a decrease in frequency is accompanied by reduction in speed. This provides the basis for the fact that the amplitude of movement in Fig. 2.20a – just as in many other charac-

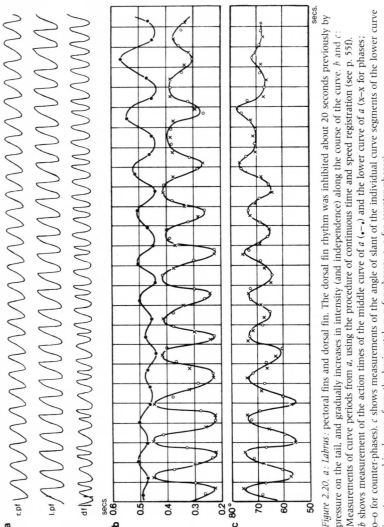

Figure 2.20. a : Labrus: pectoral fins and dorsal fin. The dorsal fin rhythm was inhibited about 20 seconds previously by pressure on the tail, and gradually increases in intensity (and independence) along the course of the curve. *b* and *c:* Measurements of curve periods from *a*, using the procedure of continuous time and speed registration (see p. 55f). *b* shows measurement of the action times of the middle curve of *a* (•–•), and the lower curve of *a* (x–x for phases; o–o for counter-phases). *c* shows measurements of the angle of slant of the individual curve segments of the lower curve in *a* (expressed in degrees from the horizontal; x–x for phases; o–o for counter-phases).

teristic cases – remains predominantly constant despite the high variation in frequency. 2. The length of the period in the rhythm exhibiting periodicity is a function of the frequency relationship between this and the other rhythm, which may be uniform or (as discussed later on) similarly periodic. This once again indicates that such periodicity must itself be based on a relationship between the two rhythms.

The method of time and speed tables. The following evaluation procedure is probably the most suitable for most cases. The method can

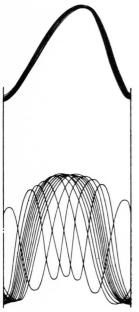

Figure 2.21. The second and third curves from Fig. 2.1a (left) have been divided into fragments of equal width and stacked one on top of the other such that the oscillations of the second curve are all concurrent (upper band). The corresponding lines of the lower curve then give the symmetrical picture shown in the lower band. (Somewhat schematized.)

be initially explained with a simple experiment. A trace of some length (something like that in Fig. 2.1a) is plotted in magnified form on transparent paper. Then, some recurrent point (e.g. the lower point of inflection of the uniform curve) is selected, and in every case a vertical line is drawn. The entire trace is then cut up along these lines into pieces of similar width, each of which contains a corresponding complete oscillation of the uniform curve and varying

57

segments of the periodic curve. When these pieces are stacked one on top of the other, so that the segments of the upper, uniform rhythm cover one another as far as possible, one obtains the picture seen in Fig. 2.21.

The first thing to be seen is that the lines of the curve lie very close to one another in some places, whilst in others they are more widely separated. This picture expresses the same statistical principle as that already outlined above in relation to certain selected points of the curve. It can also be seen that the lines of the curve do not intersect at random, but become increasingly parallel, the closer they are together. In consequence, one can arrive at an illustrative procedure if – instead of taking the lines of the curve themselves – time and speed values taken from each individual line (in the same pattern of spatial relationships as shown in Fig. 2.21) are plotted on a system of coordinates. The abscissa would represent the *position* of the individual curve inflection points (upper and lower), and the ordinate would represent the corresponding time and speed values (expressed either in seconds or cm/s or – in preference, since absolute data are not of direct interest – in percentage deviation up or down from the appropriate mean value). The times and the corresponding speeds must be recorded separately, since they can differ in behaviour. For the same reason, all times or speeds for oscillation phases in one direction and those for phases in the opposite direction must be plotted separately (or given opposing signs). In this way, one arrives at *time and speed tables* which give a summarized statement of the way in which the action time and speed of the periodic rhythm alters with change in its phase relationship to the uniform rhythm. Fig. 2.22 provides a typical example of a time and speed table for a periodic curve of the type in Fig. 2.1. From this, one can recognize the following facts: The action times for all phases (o) and all counter-phases (x) fall into roughly sinusoid lines which are mirror images of one another. The same applies to the speeds. In the same way, the time and speed curves of the same, corresponding phase directions have a mirror-image relationship. These reciprocal relationships apply to all curves of this type.

This leads on to the following rules: 1. Both the frequency and the action speed of the periodic rhythm are increased or decreased according to the momentary phase relationship to the uniform rhythm. 2. Increase in frequency always parallels an increase in

speed. 3. If a given action component of the uniform rhythm accelerates the periodic rhythm when it temporally coincides with the latter, then coincidence of this component with the counter-phase of
the periodic rhythm retards it (and vice versa). The phase and

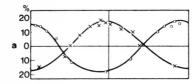
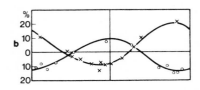

Figure 2.22. Time table (a) and speed table (b) from a trace adjacent to the specimen
given in Fig. 2.1a. The abscissa in (a) and (b) (analogous to the distance s–s in
Fig. 2.17a and b or the distance indicated by the vertical lines in Fig. 2.21)
indicates the range of possible phase relationships. The ordinate line through o
gives the level of the mean values of the times (a) and the movement speeds (b).
The individual values for the times and the speeds lie partially above and partially
below the mean value; their magnitudes are entered as % deviations from the
mean value. The time and speed values differ for the phases and counter-phases
(indicated by o and x).

In comparison with other time and speed tables (see Figs 2.34 and 2.52) it must
be noted that only the relative relationships of the four curves produced are
involved, and we are not interested here in how these four curves are limited on the
right and the left of the table and are correspondingly extracted, since the latter
depends exclusively on the selection of the reference point in the uniform rhythm.

counter-phase of the periodic rhythm, when related to the same
phase of the other rhythm, behave in opposing ways.

The time and speed table procedure can be applied anywhere
where the periodic rhythm is faster than the other rhythm, and
where the latter is completely uniform and the curve segment is
regular. An additional limitation is that whole number frequency
relationships are not permissible. (In such cases, one can only determine individual points and not the overall course of the time or
speed curves.) This method permits more detailed determination of
the extent of the periodic time and speed parameters and allows
comparison of even quite widely different forms of periodicity.

The procedure is, above all, inapplicable where the degree or form
of the periodicity are continuously altering, since the individual
points would then be scattered to varying extents around the time
and speed curves.

Conclusions from evaluation of curves

If one considers comparatively the three described methods of evaluation and their general results, they all agree in showing in some way that the periodic rhythm is influenced by the uniform one. The statistical procedure indicates the goal of this influence (approximation to the condition of absolute coordination) and the two other methods demonstrate the means through which it operates (influence exerted on the action times and speeds). If there were no such influence of one rhythm on another, neither the statistical rule of organization nor the periodic variation in the times and speeds in dependence on the frequency relationship of the rhythms, nor even any curve form in the time and speed tables, could emerge. Accordingly, we can reliably conclude that there is an influence which finds its source where one central rhythmic process originates and which exerts some kind of modifying effect on the other, simultaneously

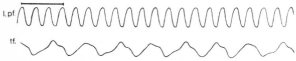

Figure 2.23. Sargus: left pectoral fin and tail fin. There is weak relative coordination (cf. Fig. 2.24).

proceeding, central rhythm. As is shown by the time and speed tables, this modification affects frequency and speed of movement in the same way. These are influenced in a positive or negative sense according to coincidence of the rhythms. This deduced influence is perhaps most immediately obvious from curves where the influenced rhythm is considerably slower than the other rhythm. Fig. 2.23 provides an example of this: the lower curve possesses a more or less uniform frequency and amplitude. Evaluation with one of the three methods described would not give a clear picture in this case. There is only a difference in the curve *form*, in that the speed of movement slightly (but distinctly) oscillates up and down in the tempo of the upper rhythm. The relationship between this variation and the uniform rhythm is elucidated when the trace extract is treated in a manner analogous to that applied to Fig. 2.1 in Fig. 2.21 (i.e. when it is fragmented at intervals dictated by the frequency of the upper rhythm and the pieces are then superimposed).

60

A comparison of the band in Fig. 2.24a with that in b, in which the lower rhythm performed quite even and sinusoid oscillations, shows a characteristic difference: with the influenced oscillation band there are two regions, one of which (1 : left) is marked by the fact that all upward lines are steeper than in the normal comparison band, while all downward lines are flatter. In the other region (2 : right), the converse is the case. Thus, one can see here how (1) the phases and counter-phases are subject to contrasting influences and (2) this influence itself is inversed as the phase of the effective rhythm is changed. This mode of influence of one central rhythm

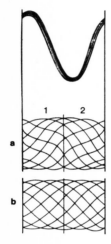

Figure 2.24. *a*: the curve from Fig. 2.23 is fragmented and the pieces superimposed, such that the oscillations of the upper rhythm are concurrent (analogous to the treatment in Fig. 2.21). From the lower curve of Fig. 2.23, one then obtains the lower band in 2.24a. *b*: a similarly evaluated trace segment from the same fish, for comparison, where the tail fin rhythm was quite uniform and uninfluenced (somewhat schematized).

on another is of decisive importance. It is always found in the same way in all forms of periodicity of the type considered, however much they may differ from one another. The additional phenomenon of frequency variation can be derived from this as a secondary effect. This is all based on the fact that a phase which has been accelerated (retarded) is concluded correspondingly earlier (later) and that the following counter-phase accordingly begins earlier (later).

It remains to be seen (1) in which phase relationships acceleration and retardation of the action of the dependent rhythm occur, and (2) whether the extent of this effect is constant or whether the intensity (*ceteris paribus*) varies according to the momentary relationships between the rhythms. With respect to the first point, comparison of numerous curves demonstrates that there is no unitary answer to the question concerning the phase relationships associated with acceleration and retardation. The answer varies according to the rhythms concerned or according to the curves plotted for different fin rays utilized. However, for all cases without exception it is a rule *that the dependent rhythm is retarded as soon as it is ahead of the phase position maintained in absolute coordination, and is (conversely) accelerated as soon as it is behind this phase position.* Now, since the dependent rhythm — whether it has a higher or a lower inherent frequency than the other rhythm — is smoothly displaced towards the latter and exhibits in sequence all possible phase relationships to this rhythm, it is periodically accelerated and retarded. This applies equally to relatively faster and to relatively slower rhythms. Thus, it is not the case that one rhythm which is slower is merely accelerated by the influencing rhythm, whilst a faster one is retarded. Quite independently of this, the momentary influence is decided by the prevailing phase relationship.

As far as the second point is concerned (i.e. *intensity of the influence*, as determined from the momentary degree of departure of the action times and speeds from the uninfluenced norm, or overall mean value), the following statement applies: As soon as the two rhythms maintain the mutual phase relationship which is appropriate to absolute coordination, the influence of one rhythm on the other is zero. If the influenced rhythm gradually departs from this spatial relationship by retardation or acceleration, then the degree of positive or negative influence gradually increases to reach a maximum, which lies at a deviation of roughly one-quarter of a complete oscillation from the position of absolute coordination. As the deviation increases further, the influence begins to decrease, finally returning to zero at a displacement of half of a complete oscillation from the absolute coordination position. Thus there are two spatial relationships in which there is no demonstrable influence. The first is characterized by the fact that any departure from the relationship as a result of acceleration or retardation of one rhythm immediately evokes a (retarding or accelerating) force opposing this tendency from the other rhythm. Therefore, in most cases this spatial relationship represents a *stable equilibrium* — that which is referred to as 'absolute' coordination. On the other hand, the second spatial relationship, in which there is no influence exerted, is a *labile* condition. From this position, any acceleration of the influenced rhythm evokes a further accelerating force from the independent rhythm, and retardation evokes a further retarding force.

This fact, i.e. that the intensity of the effect increases from zero to a certain level and then falls away again, has the consequence that even with absolute coordination the phase relationships between the two rhythms are not completely fixed but can vary in smooth transition within a certain range. This emerges more or less clearly in all cases where the two rhythms have noticeably different inherent frequencies. The faster rhythm is then always in advance to a specific, constant extent (and vice versa), so that when there is absolute coordination one can predict which of the two rhythms will be faster or slower when the other rhythm is arrested.

The results so far can be summarized thus: The various rhythmic processes in the spinal cord of the fish, which give rise to the oscillations of the individual fins, are fundamentally able to proceed quite independently of one another and with quite different frequencies. Each of these rhythms, as long as the external and internal conditions remain constant, will tend to maintain its inherent frequency. We can refer to this as the *maintenance tendency*. As a further phenomenon, there is very often emergence of a tendency for one rhythm to impose its inherent frequency on the other rhythm. This is achieved in that one rhythm – the *dominant* one – retards the performance of the other as long as it is in advance of the requisite phase position (the *coaction position*),[1] and accelerates it as long as it is behind the coaction position. I have provisionally suggested the term *magnet effect* (M-effect) for this influence of one rhythm on another (a term which is not intended to indicate any supposition about the physical mechanism involved!). The magnet effect and the maintenance tendency of the dependent rhythm are in mutual opposition. If the former predominates, then there is continuous agreement in frequency under absolute coordination; if the latter predominates, there is relative coordination – the frequencies of the rhythms differ, and the dependent rhythm, under the magnet effect of the dominant rhythm, exhibits periodicity whose form is determined by the reciprocal frequency relationship and whose extent is governed by the intensity of the magnet effect. Thus, the M-effect – stated briefly – is a factor which provides 'associations' in the

[1] This term is used since the coordinated elements exhibit the tendency to operate together – to *coact*.

truest sense of the word between different central processes, which are in principle independent.

Reciprocal frequency influence between two rhythms

In the simple case examined thus far, one rhythm has been dominant and the other dependent. More often, the situation is that two or more rhythms exert a mutual magnet effect upon one another. An example of this has already been given in Fig. 2.20a, as is shown in the evaluation in Fig. 2.20b. For all such cases there emerges a generally valid principle: At any one time, two rhythms which influence one another do so *in opposition*. If, in a particular phase relationship, the first rhythm has an accelerating effect on the second, then the latter has a retarding effect upon the former. In the periodic frequency oscillations which are produced, a frequency maximum of one rhythm is thus always correlated with a minimum in the other rhythm and vice versa (see Fig. 2.20b). This rule is nothing more than an expression of the fact that the phase relationship in which one rhythm (a) endeavours to maintain another rhythm (b) is the same as that to which (b) also endeavours to draw (a). Through the fact that (b) and (a) exert mutual effects (one accelerating and one retarding) in any position departing from that of coaction, their forces summate in tending towards the coactive relationship. Accordingly, we can establish that *the coactive relationship towards which two rhythms tend is the same for both rhythms.*

The intensity relationship in mutual magnet effects does not follow a set rule. This can vary so widely that a rhythm which is completely dominant under certain experimental conditions will, under other conditions, become completely dependent upon the rhythm which it previously dominated. All stimuli and other influences (alteration of the water supply, of the temperature, and so on) which generally fortify a rhythm – as is most obvious in greater amplitude of oscillation and in increasing independence of influence from other rhythms – also tend to produce a corresponding increase in its M-effect on other rhythms. (That is, as long as this effect is already present. One can, of course, equally well find a strongly active rhythm with no M-effect.) The specimen trace in Fig. 2.20 provides an example of this phenomenon as well.

A further, generally applicable rule emerges automatically from the laws outlined above. If, in cases of mutual M-influences, absolute coordination is produced, the frequency which they jointly adopt lies *between* the two individual frequencies which they had previously maintained. The nearer the joint frequency is to the prior individual frequency of one of the two rhythms, the greater was the relative M-effect of that rhythm. Thus, the stable condition which emerges is a resultant of two force components, which may be equal or unequal to varying degrees.

This leads on to a further question. After attainment of absolute coordination, is the tendency of the two rhythms to maintain their previous individual frequencies now eliminated, or perhaps converted to a tendency for maintenance of the 'compromise' frequency? The answer to this must be 'no'. For, in every case of this kind, when one of the two rhythms spontaneously ceases, or is inhibited or sufficiently weakened by a stimulus, the other rhythm at once exhibits precisely the individual frequency which it possessed previously (Fig. 2.16). From this we can conclude that the 'struggle' between the M-effect and the maintenance tendency continues *in latent form* even after attainment of absolute coordination.

Special cases of the magnet effect

In a number of cases, the relationship between the rhythms can assume an appearance differing somewhat from the normal picture, but without presenting any difficulty for derivation of the observed traces from the operation of the same central process. Instead of simply accelerating or retarding, as in the cases so far described, one rhythm may 'push' another – as yet inactive – rhythm over the threshold, or it may 'suppress' the activity of another rhythm. These cases will now be considered.

Effect of individual rhythmic actions upon one another. With rhythms of the 'all-or-nothing' type, it is characteristic that following long pauses (after operation, on emergence from narcosis, etc.) they are frequently irregular in activity in the initial period, only exhibiting isolated actions with intervening intervals. Even between isolated rhythms of this kind, one can usually demonstrate either relative or absolute coordination. In the former case, extensive traces show that the two rhythms involved do in fact emerge and disappear quite independently of one another; but statistical evaluation demonstrates that there is predominance of actions in which both are per-

The Behavioural Physiology of Animals and Man

formed simultaneously, in the coactive position. In the latter case (an example of which is given in Fig. 2.25 a–c), the relationship is even more rigid: whenever the two rhythms are simultaneously active, they are coactive from the outset. This comes about in that one of the two rhythms begins at a precise time after the second, such that the coactive position is maintained from the beginning. Thus, the rhythm which emerges first 'ensures' that the other rhythm does not emerge at random, but at a quite specific time.

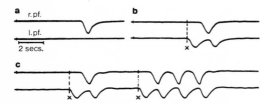

Figure 2.25. *Labrus*: extracts from a lengthy trace in which the pectoral fin movements were recorded in the post-operative phase. The two rhythms sometimes appear in isolation and sometimes together, with the intervening intervals gradually becoming shorter. (x marks the instant of activity-onset for the left pectoral fin.)

Relationship between a continuous rhythm and isolated oscillations. In cases where one central rhythm is continuously active whilst another exhibits isolated oscillations, similar relationships are again observed. As far as the intermittent rhythm is concerned, one can observe that it does not appear at random moments, but that appearances are clustered or exclusively arranged such that the emergent activity is coactive with that of the continuous rhythm (example in Fig. 2.14b). Thus, in this case too the continuous rhythm determines the moment of onset. Conversely, the individual actions often take effect on the simultaneous oscillation of the other rhythm when the two rhythms have different speeds. In the example given in Fig. 2.26, any single oscillation of the lower rhythm which temporally co-incides with a single pectoral fin action is drawn out by more than a quarter in length.

From these examples it is evident that one proceeding central rhythm can have an effect upon another when the latter is not yet active, but is presumably not far from the threshold of initiation. Here, the magnet effect has a different appearance. Instead of the usual effect of accelerating an active process, there is a priming effect of elevation above the threshold. Instead of retardation, there is retention below the threshold of initiation. In principle, these two

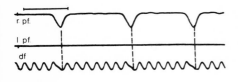

Figure 2.26. Labrus: pectoral fins and dorsal fin. The lower rhythm is continuously active, whilst the upper occasionally exhibits a single action (see reference lines), which always has a specific spatial relationship to the lower rhythm and retards its performance.

effects exerted by the magnet influence are probably equivalent. It is quite natural that an influence which pushes a process over a threshold (or lowers the threshold concerned) will have an accelerating effect when applied to the ongoing process. Conversely, a retarding effect, following resultant arrest of the process, will continue to raise the threshold of initiation.

Suppression of actions. Fig. 2.27 provides an example of a case which is not at all uncommon, where an action of the dependent rhythm

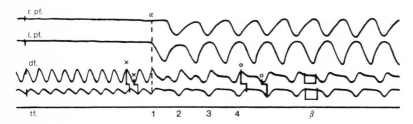

Figure 2.27. Labrus: pectoral fins, dorsal fin (recording of the movement of two fin rays) and tail fin. At α, there is spontaneous emergence of the pectoral fin movements. (The reference lines at x . . . x and o . . . o are intended to indicate lengthening of the phase distance of the movement of the two dorsal fin rays following transition to the slow pectoral rhythm.) See text for explanation of β.

does occur 'at the wrong time', but − as a result of the M-effect of the dominant rhythm − cannot achieve full realization.

Up to the point α, only the dorsal fin rhythm is active; subsequently, the two pectoral fin rhythms emerge with a tempo which is more than twice as slow. Despite the obvious presence of a strong M-effect, the dorsal fin rhythm does not at once follow this tempo. Instead (between 1 and 3), there is a small intervening action, such that the third curve temporarily presents a picture of an *'Alternans'*. This greatly limited 'intervening beat' is completely supressed in the subsequent activity. In its place, there is a component of marked retardation of action (marked with β at one point).

Fig. 2.28 provides another example showing an analogous effect. The upper, previously inhibited rhythm (as in Fig. 2.19a) initially completely omits every second action. Thereafter (at the points

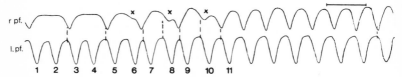

Figure 2.28. Labrus: the right pectoral fin was inhibited just beforehand (as in the experiment of Fig. 2.19), and it is in the process of gradually attaining its previous frequency and intensity of action. At the points marked x, there are emergent actions which are not coactive with the other rhythm, and they are all 'suppressed' by the lower rhythm to varying extents, according to their intensity.

marked x above 6 and 8), there is temporary pronounced suppression of action. At 10, the corresponding action is already much stronger, and from 11 onwards there is absolute coordination.

In this case, the dominant lower rhythm is also affected to some extent. At the points marked x, there is in each case relatively early appearance of one action of the dependent rhythm, which attains its peak in advance despite the strong opposition (cf. the reference lines drawn between the two rhythms). Each time, the lower rhythm is forced to accelerate, such that the curve troughs 6 and 7, and 8 and 9 are closer together.

These examples show that a hampering magnet effect can become so strong that, instead of mere retardation of an action, there can be actual premature arrest. In such cases, the M-effect operates against the maintenance tendency and the all-or-nothing tendency of the dependent rhythm and imposes premature directional change in the latter. A reaction of this kind is common when the dependent rhythm has already been weakened by other effects (as in Fig. 2.28, where there was prior inhibition). Otherwise, it is completely lacking with some rhythms, whilst with others it is a fairly regular occurrence when the M-effect is strong. The individual strongly reduced amplitudes in Fig. 2.16 (right: 'intervening beats') are to be interpreted in this way.

Model experiments

Since the described effects within the CNS have not been recorded previously, and since they cannot be easily traced to anything which is already known, it is justifiable to explain the introduced ideas and concepts using a physical model of minimal complexity, which can produce movement pictures which are *formally* in agreement with all the important points (although its *mechanism* definitely has nothing in common with that in the central nervous system). The apparatus illustrated in Fig. 2.29II mainly consists of (1) a pendulum which rhythmically moves a viscous mass (syrup) backwards and forwards, and (2) an axle rotated by a weight and dipping into the viscous fluid, such that the frictional resistance of an eccentric sphere (e.g.) at its tip determines the speed of rotation of the axis in the fluid. As long as the pendulum is immobile, the axis rotates evenly, and the lever (H_2) which it operates oscillates in sinusoidal fashion. As soon as the pendulum is oscillated, the sphere in the syrup is accelerated or retarded, according to whether the pendulum is momentarily swinging in the same direction or the opposite direction. The even rotation is replaced by periodic rotation. The form and the extent of the periodicity depend upon the frequency relationship between the two rhythms and the amplitude of the pendulum arc. With this simple apparatus, one can formally replicate all of the main phenomena associated with the M-effect, along with all of the novel concepts associated with it.

As long as the pendulum is immobile, the rotating axis possesses a specific 'inherent frequency', which is determined by the relationship between the driving weight and the frictional resistance of the sphere in the syrup. The same inherent frequency can be maintained with a light weight and a small sphere, or a heavy weight and a large sphere. In the former case, a given supplementary driving or inhibiting force would have a big influence; in the latter case, the influence would be small. Thus, with the same frequency, the 'maintenance tendency' varies in this instance. The frequency itself depends upon the speed of performance of the individual phases ('action speed'), any alteration of which automatically corresponds to a change in frequency.

The variation in speed – and thus in frequency – brought about by the oscillation of the pendulum is formally quite analogous to the M-effect. The correspondence extends to the following features: (1) Oscillation of the pendulum with a large amplitude imposes upon the 'dependent' rhythm the frequency of the pendulum, with a quite specific phase relationship ('absolute coordination'; Fig. 2.30a right). (2) Below a certain amplitude of the pendulum, the dependent

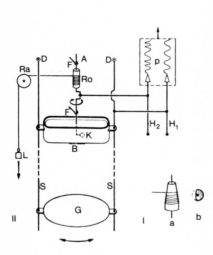

Figure 2.29 II: sketch of the mechanical model with which the curves in Figs 2.30, 2.31 and 2.32 were prepared. It consists of two bars, freely suspended from the points D, which are loaded with a weight (G) below and carry a reservoir (B) containing a viscous fluid (syrup). Oscillation of this pendulum system (see arrow beneath the weight) moves the recording lever H_1 to and fro. The axle A rotates around a fixed point (F) and is driven by a thread which is wound round the cylinder R_0 and is linked to a weight (L) over a pulley (R_a). The lower end of the axle A, which operates the recording lever H_2, bears an eccentric sphere (K) dipping into the reservoir (B). (See text for the functioning of this apparatus.)
Ia: conical driving cylinder.
Ib: transverse section through a driving cylinder which is flattened on one side.

rhythm does not conform with the 'dominant' pendulum rhythm but is only subject to periodic increases and decreases in speed following the latter's tempo (Fig. 2.30a left). (3) The length of the periods varies and depends upon the reciprocal frequency relationship (cf. Fig. 2.30a and b). Evaluation of such periods using one of the three procedures described above demonstrates that the observed behaviour exactly corresponds to that in periodic fin movements.

In cases where the dependent rhythm is the slower of the two, modification of the frequency becomes reduced with respect to variation in the speed of the activity (cf. Figs. 2.31 and 2.23). (4) Transition from absolute to relative coordination can be achieved in two ways: either through weakening of the 'M-effect' (= amplitude of the pendulum), as shown in Fig. 2.30c, or through increase (or retardation) of the latent 'inherent frequency' of the dependent rhythm, using a

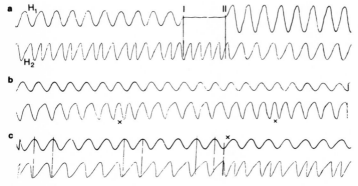

Figure 2.30. Model curves originating from the apparatus depicted in Fig. 2.29. Above: curve for the pendulum H_1; below: rotation of the axle H_2. *a:* There is 'relative coordination' up to I; between I and II, the pendulum is held still and the 'dependent' rhythm continues with its own 'inherent frequency'. After II, there is oscillation of the pendulum with greater amplitude and conformity of the 'dependent' rhythm (= 'absolute coordination'). *b:* Relative coordination with long periods (frequency maxima marked by x). The inherent frequencies of the pendulum and the axle rotation differ only slightly. *c:* Abrupt transition from 'absolute' to 'relative' coordination (x), with slow restriction of the pendulum amplitude (= 'M-effect'). With the aid of the reference lines, one can see that the phrases of the two rhythms have gradually become somewhat out of step even before x (acceleration of lower rhythm).

Figure 2.31. Model curve as in Fig. 2.30. 'Relative coordination' in which the dependent rhythm is considerably slower than the dominant one (cf. Fig. 2.23).

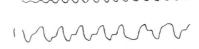

conical driving cylinder (Fig. 2.29 Ia). The converse applies for the transition from relative to absolute coordination. (5) The critical point at which one kind of coordination is converted to another is not dependent upon the relationship between the 'M-effect' and the 'maintenance tendency'; instead, it is similarly dependent upon the relationship between the frequencies of the two rhythms. The greater the difference in frequency, *ceteris paribus*, the greater must be the

M-effect, if 'absolute coordination' is to emerge. (6) In a case of 'absolute coordination' there is no rigid maintenance of a specific phase position (the requisite coactive position). Just as with the fish rhythms, the dependent rhythm – if it has a faster inherent frequency – is continually a little in advance of the 'coactive position' (up to a maximum of a quarter of a cycle), and the rhythm with the slower inherent frequency always remains behind to a greater or lesser extent. Only when the 'M-effect' is very strong, or the 'main-

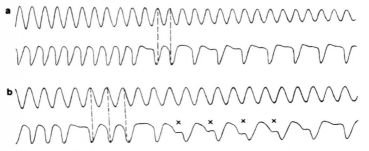

Figure 2.32. Model curves as in Figs 2.30 and 2.31. The drive of the rotating axis (lower curve) was in this case provided by the flattened cylinder shown in Fig. 2.29 Ib in order to produce the 'all-or-nothing' response. In *a* and *b*, the inherent frequency of the 'dependent' rhythm is lower than that of the dominant one. There is transition to a 1:2 relationship with reduction of the pendulum amplitude (= 'M-effect'). The picture in 2.32a (right) is analogous to that in 2.19a, whilst that in 2.32b (right) resembles Fig. 2.28 in appearance (cf. in particular the points marked x). The reference lines in 2.32a and b emphasize the fact that, at the points concerned, the curve peaks of the 'dependent' rhythm are farther from one another than those of the dominant rhythm. This resembles the situation as seen in Fig. 2.19b, c.

tenance tendency' very weak, does absolute coordination produce the exact 'coactive position'.

In all of the points concerned, the spinal cord of the fish and the model are in formal agreement. Over and above this, one can (for example) produce the special cases illustrated in Figs 2.19 and 2.28, using the model. This can be achieved by imitating in the model the peculiarity of the all-or-none tendency present in many fish rhythms, using a driving cylinder flattened on one side (Fig. 2.29 Ib) to reduce the drive of the rotating axle at one point – the 'resting position' – almost to zero. The 'rhythm' operated in this way consequently has a tendency to incorporate intervals at this point. The two sample curves in Fig. 2.32 show images which are quite characteristic for rhythms cf the all-or-nothing type. In both cases, the inherent frequency of the dependent rhythm is below that of the pendulum frequency. When the 'M-effect' declines (= reduction in pendulum amplitude), the absolute coordination first evident in Fig. 2.32a gives way to a 2 : 1 relationship, as seen (for example) in Fig. 2.19a for fish rhythms. In Fig. 2.32b, following two isolated omissions of actions, one can see a series of typical incomplete 'suppressed' actions which are quite equivalent to the oscillations of similar form seen in fish rhythms under comparable conditions (cf. Fig. 2.32b with Fig. 2.28).

The examples cited clearly show that the model can replicate the natural relationships quite as far as one can reasonably expect, considering the differences between the two mechanisms. From this fact one can draw two conclusions: (1) The description which has been given of periodicity in fish rhythms of the type considered, together with the associated concepts introduced, very probably provide a *representation* (*not* an interpretation or an explanation!) of the production of periodicity.[1] (2) The coupling effect in the CNS based on the M-effect is clearly, and to a large extent undistortedly, evident in the fin movements.

One might be tempted by the model to replace the term 'M-effect' by 'coupling effect'; but this does not seem to me to be advisable. In physics (and probably in neurophysiology as well) there are numerous forms of coupling, and in this case we are concerned with coupling of a quite specific kind, which could be termed 'mechanical coupling through a viscous medium' in the case of the model. This form of coupling, which – as far as I know – has not yet been the object of detailed physical analysis, is fundamentally distinct from other forms of coupling of rhythmic processes. The term 'coupling' would therefore be too general as a substitute for 'M-effect'. Thus, the old term should be retained as long as the physical process in the CNS is not more fully understood and as long as one does not find a process in the CNS which literally deserves the term 'magnetic' (rather than meriting this purely descriptively, as in the case of the M-effect).

---

[1] It should be mentioned that the information derived about the M-effect was not acquired using the knowledge gleaned from the model, since the latter was constructed at a much later stage.

SUPERIMPOSITION PHENOMENA

General rules

The second group of phenomena seen with fin rhythms includes cases of mutual influence where any tendency for one rhythm to impose its operating tempo on the other is completely lacking. The cyclical tracings of this type are clearly distinct from the previous type even in their external appearance. The tracings closely resemble the well-known physicists' patterns showing superimposition (interference) of oscillations of different frequencies. Examples are given in Figs. 2.2, 2.3 and 2.33.

For all such patterns of periodicity the following rules always apply: (1) They only emerge when there are at least two rhythms of differing frequency operating simultaneously. (2) The rhythm determining the periodicity retains its previous amplitude as a *mean amplitude*; both parameters oscillate in a periodic manner around these mean values. (3) Even when these periodic oscillations attain a maximum, there is never a resulting attachment of one rhythm to another. (4) The length of the resultant periods depends upon the relationship between the frequencies of the two rhythms. If the periodic rhythm is considerably faster, the period length is determined by the tempo of the uniform rhythm (Fig. 2.3); the same applies if the former is considerably slower. In all other cases, a new period begins when one rhythm has overtaken the other to the extent of one oscillation (Figs 2.2, 2.33). (5) Exact measuring of the traces demonstrates, for each of the three processes described above, behaviour which is characteristically distinct from the cases involving the *magnet effect*: (a) Statistical evaluation of the relationship between two particular recurring points on the trace does indeed give a shallow variation curve similar to that in Fig. 2.17, but the *maximum* is located exactly where one finds the *minimum* in the case of an M-effect between the two rhythms. (b) When the times and speeds are continuously evaluated, both parameters exhibit periodic variation; however, the two curves obtained are not mirror images of one another (as with the M-effect – see Fig. 2.20b and c), but parallel, as can be seen in Fig. 2.33b. (c) Evaluation of a trace segment in the form of time and speed tables (see Fig. 2.34) gives two time curves which are mirror images of one another, and two similar speed curves. However, in contrast to the case of an

M-effect, each of the time curves shows a course parallel to that of the correlated speed curve. These regular effects permit two reliable conclusions: 1. These cases of periodicity are also based upon

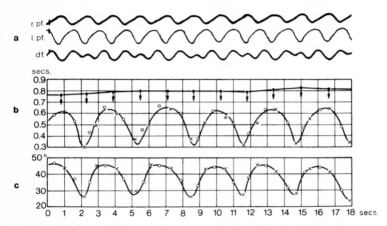

Figure 2.33. Labrus: relative coordination. *a:* The dorsal fin rhythm provides an example of a pure superimposition trace. In *b* and *c*, the lower curve of *a* is continuously evaluated in the same way as described for Fig. 2.20. In *b*, the upper (almost straight) trace represents the timings for one of the non-periodic pectoral fin rhythms from *a*. (The arrows indicate the intervals at which one would have to section the trace of *a* and superimpose the segments in order to achieve a regular line diagram analogous to that in Fig. 2.21.)

the influence of the uniform rhythm upon the periodic one. 2. The influence in question must be fundamentally different from the M-effect.

A working hypothesis and experiments to test it

The extensive similarity to superimposed oscillations leads to the *hypothesis* that one central rhythmic process is in some way superimposed[1] on the other, or that the two processes contact one another so intimately that the elements conducting excitation to the musculature respond simultaneously to the two rhythmic processes, such that

---

[1] One is probably particularly tempted by these traces to apply a harmonic analysis using the Fourier method. This is not advisable, since the two principal oscillations incorporated in the observed periodicity are in any case apparent and the possible identification of further, more rapid, oscillations (which could be based on the recording process or on the imperfectly sinusoidal beat action of the fins) would only lead to erroneous conclusions.

75

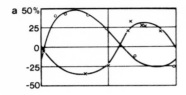 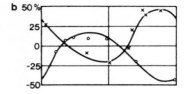

Figure 2.34. Time (*a*) and speed (*b*) tables for the lower trace of Fig. 2.33a. For details, see Fig. 2.22.

the superimposition resultant of the two processes (i.e. their summation at any point in time) would be expressed in the action of the muscles.

The actual close correspondence between all of these oscillation traces for fish fins and physical interference curves can easily be demonstrated with experimental models. It is quite easy to replicate any fin periodicity with a model trace showing formal correspondence. Fig. 2.35 provides an example of interference between two mechanical pendulum oscillations. The middle, periodic curve is evaluated below (Fig. 2.35b, c), using the method of continuous time and speed recording. When this is compared with Fig. 2.33b, c, one can see that the resulting time and speed curves exhibit parallel courses in both cases and that they all correspond in having the shape of an inverted garland.

This *superimposition hypothesis* leads on to a number of corollaries through which the applicability of the hypothesis can be tested.

1. If the hypothesis is correct, in a case where two different rhythms are superimposed on a third, their superimposition effect should be *additive* at a particular phase relationship between the two and *subtractive* (i.e. mutually cancelling) with the converse phase relationship. This requirement is met; Fig. 2.36 provides an example: At the points I and III, the effect of the two upper rhythms is added to the third, which exhibits pronounced 'periodicity' (*'Alternans'*). At the point II, where the two rhythms exhibit a phase relationship opposite to that at points I and III, the lower rhythm becomes temporarily completely uniform.

2. When the superimposition effect of one rhythm increases beyond a certain level, at certain frequency relationships it must produce in the rhythm subject to superimposition an action in a direction

opposite to that usually found. The schema in Fig. 2.37a, b explains this for a frequency relationship of 1 : 2. When there is weak superimposition (Fig. 2.37a), the amplitude and the fre-

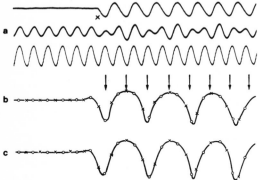

Figure 2.35. a: an experimental model. The movements of two mechanical pendulums attached to the same axis (upper and lower traces) are transmitted through rubber threads to a light lever resting on the axis between them (middle trace). The lever indicates their superimposition (i.e. the sum of the two pendulum movements divided by two). One pendulum is set in motion from the point x onwards. *b* and *c:* evaluation of the superimposition trace, in the same way as for Fig. 2.33b and c. (Actual measurements have been omitted since they are unimportant in an experimental model of this kind.)

quency vary in alternation (cf. the lower curve in Fig. 2.36); when superimposition is strong (Fig. 2.37b), the superimposition trace must exhibit omission of every second action of the rapid rhythm and replacement of the short upward peak by a blunt, downward projection (i.e. movement in the opposite direction). Fig. 2.38 provides an example trace: as a result of continuous alteration of the experimental conditions (by respiratory water restriction), the recording trace is extremely variable. From 'a' onwards, the frequency relationship is 1 : 2. One can see that the lower rhythm passes over to an '*Alternans*' of increasing clarity. The small trace peak gradually diminishes in size until it eventually gives way

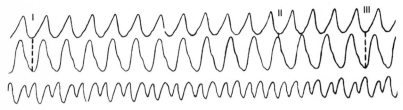

Figure 2.36. Labrus: pectoral fins and dorsal fin (below). Example of interference of the two pectoral fin rhythms in their effect upon the dorsal fin rhythm. (See text for details.)

(points marked x) to a bulge in the opposite direction, in a manner analogous to that seen in the schema of Fig. 2.37b.

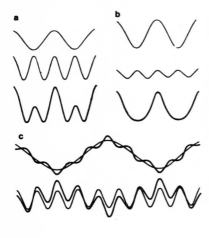

Figure 2.37. Schematic diagrams to explain different superimposition traces. The sinus oscillations are represented by thin lines and their additive product by a thick line. In *a* and *b*, there is superimposition of two oscillations with a frequency relationship of 1 : 2; in *a*, the more rapid rhythm has the higher amplitude, in *b* the slower rhythm (cf. Fig. 2.38, right). In *c*, there is mutual superimposition of two oscillations with quite different frequencies. The traces are drawn with a mirror-image relationship to one another, and they are reduced to half their amplitude when added to one another in order to provide a better comparison with the tracing shown in Fig. 2.39, right (between - and ∶).

3. The hypothesis must also apply to cases where the two rhythms exert reciprocal effects. Here, each rhythm – apart from maintaining its inherent frequency – must also incorporate the tempo of the partner rhythm. Fig. 2.39 provides an example of this for a case where, as in Fig. 2.38, the frequency and amplitude of both

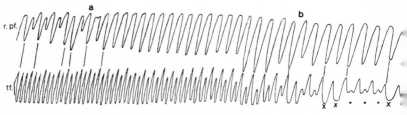

Figure 2.38. Sargus: right pectoral fin and tail fin. Change in coordination induced by respiratory water restriction. Before the point a, the upper rhythm is exclusively influenced by the lower; later, the opposite applies, and the lower is dependent upon the upper. Before a, the frequency relationship is (predominantly) 2 : 3, above it is a 3 : 2 '*Alternans*' (cf. the schema in Fig. 2.55, curve 3C). After a, there is a 1 : 2 '*Alternans*' of increasing clarity in the lower trace (for the points marked ● and x, cf. schemata in Fig. 2.37a and b).

Figure 2.39. Serranus: right pectoral fin and tail fin. Modification of action by water exclusion. Between x and **.** the frequency relationship is 1:2; following this there is slowing down, and at **:** there is arrest of the upper rhythm. This is an example of reciprocal superimposition (for the section **. – :** compare the schema in Fig. 2.37c).

rhythms are continuously changing as a result of respiratory water restriction. From the point x onwards, there is a 1:2 relationship; but in this case it is not only the lower rhythm which exhibits an *'Alternans'* (as in Fig. 2.38), since the upper rhythm also shows some degree of following the lower one. From the point **.** onwards, the upper rhythm gradually becomes slower, finally stopping completely at the point **:**. Here, one can see even more clearly how the slow beats are superimposed by restricted oscillations in the tempo of the lower rhythm, whilst the latter – apart from exhibiting its frequent actions – oscillates to the slow tempo of the upper rhythm. (The fact that this movement represents a mirror image of the upper one, rather than coinciding in direction, is simply a result of the lever transmission in recording.)

4. When a rhythm 'a' is superimposed on another, 'b', inhibition of 'b' (i.e. of one component) must be followed by exclusive retention of the component 'a' in both oscillations. Thus – assuming that in an experiment there is continued excitability of those elements of 'b' which respond to the superimposition process and transmit it to the muscles – the fin associated with the central rhythm 'b' should continue to beat with weak amplitude to the tempo of the rhythm 'a'. This phenomenon can also be observed in many cases. Fig. 2.14b has already provided an example of this, showing how, under these conditions, the dependent fin has two types of amplitude – a small amplitude based (according to the hypothesis) on mere 'accompanying response' to the other, distant rhythm, and a large amplitude (having its own, inherent frequency – in this case a slower one), which emerges when the fin's own, inherent central rhythm begins to function. (Fig. 2.14a, e and f, together with the second trace in Fig. 2.14b, show as a control that the restricted oscillation is not merely based upon a mechanical effect exerted by the lower rhythm.) One can aptly

refer to these movements as 'accompanying movements' and interpret the entire superimposition phenomenon as a form of activity in which there is addition, to the oscillation at the inherent frequency, of simultaneous accompanying movement at the tempo of the other rhythm.

The arguments listed, which can be expanded with a number of other points, undoubtedly fuse the concept of superimposition into a well-supported theory. So far, no observations have been made which throw doubt on the correctness of the theory. One can only obtain a clearer picture of the manner in which such central superimposition could occur by becoming closely acquainted with the relationships within a central rhythm (described below). This, in fact, raises the possibility that the superimposition phenomenon is to be understood as a special case of the magnet effect.

INTERACTION OF THE MAGNET EFFECT AND SUPERIMPOSITION

In the vast majority of cases, the two separately considered phenomena of the magnet effect and superimposition occur together; periodicity in which superimposition occurs without presence of an M-effect represents an extreme case, just as much as a situation where there is pure magnet effect without simultaneous superimposition. The intensity relationship between the M-effect and superimposition, when both are operating at once, is extremely variable; both factors can be independently influenced by stimuli or by alteration in the state of the prepared fish. Nevertheless, there is a rigidly controlled relationship between the two. The *coactive situation* towards which the M-effect operates is always that *in which the amplitude maximum occurs with superimposition*, whilst the amplitude minimum conversely occurs with the opposing phase relationship. Thus, if there is a case of combined superimposition and magnet effect leading to attachment of the dependent rhythm to the dominant one, the former is continually maintained with the phase relationship in which it reaches a maximum of amplification in action through superimposition by the dominant rhythm. Fig. 2.40 provides an example of this.

This important rule automatically leads on to further peculiarities concerning the curve shape and evaluation results when there is combined operation of the M-effect and superimposition. The middle curve of Fig. 2.40 provides a characteristic picture, which is distinct even in its external features from that occurring with pure superimposition. Whereas, in the case of pure superimposition with this frequency relationship, each individual cycle would be symmetrical in its right and left halves (cf. Fig. 2.2), in this instance the amplitude maximum is moved up quite close to the left-hand minimum and away from the right-hand one.

The evaluation of such traces with time and speed tables (e.g. the middle trace of Fig. 2.40) gives a result intermediate between that

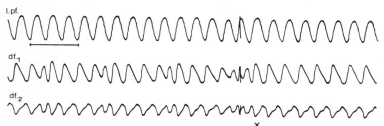

Figure 2.40. Crenilabrus: pectoral fin and two dorsal fin rays. An example of combined magnet effect and superimposition, particularly in the middle trace (right/left asymmetry in periodicity). In the lower curve, the superimposition effect is slight. After the point x, there is spontaneous conversion to absolute coordination. (See text for details.)

with a pure magnet effect and that with pure superimposition. Both time curves and both speed curves represent mirror-image pairs; however, each associated pair of time and speed curves exhibits neither an opposed (M-effect) nor a parallel (pure superimposition) relationship, but a laterally displaced situation. The degree of this displacement (or, alternatively, the degree of approximation to an exclusive M-effect or exclusive superimposition) can be used as an indicator of the intensity relationship between the two effects for a given curve form.

The co-action of the M-effect and superimposition plays a considerable rôle, particularly in cases of complex periodicity involving more than two rhythms (see p. 118 et seq.).

General data concerning the nature of central rhythmicity

The account given so far has treated the relationship between central rhythms in a *simplified* form, to the extent that all of the central elements which are involved in the movement of a fin are briefly summarized as 'a rhythm'. Therefore, before we go into the difficult question of the reciprocal relationship between the elements contained within such a rhythm, it is appropriate at this point to bring in a number of observations and experiments permitting more detailed characterization of this central rhythmic function.

AUTOMATIC ACTIVITY

There is an old and widely accepted idea that locomotor activities represent 'chain-reflexes' of a kind which always involve reflex elicitation of one motor phase by its predecessor, and that the locomotor coordination of different limbs is produced in the same way from individual, mutually elicitatory reflexes. Everything that is concerned in the purely central 'automatic' processes of fish speaks against this concept. The experiments carried out in this area (von Holst, 1935b; 1938a) are the following:

1. If all the dorsal roots of the spinal cord of a fish (tench, *Tinca*) are transected whilst leaving the ventral roots intact, such that no further responses of any kind are given to any stimuli associated with either the dorsal skin or the musculature, the rhythmic movements of the dorsal musculature are by no means suppressed; instead, well-coordinated rhythmic swimming movements occur (e.g. in response to optic or static stimuli, or even spontaneously).

2. If one takes an eel (*Anguilla*), in which the dorsal *and* ventral spinal cord roots have been transected for some distance in the middle of the body, and the fish is attached by this middle section in the water such that reciprocal mechanical influence between the head and tail sections is rendered impossible, the two ends of the body perform swimming movements with the same rhythm and with the normal phase relationship of the intact eel.

3. If one operates upon an eel and inserts a rigid splint alongside the vertebral column, so that the mobility of the middle section of the body is more or less completely suppressed, the head and tail section continue to move in a normal, coordinated fashion.

4. Analogous observations can also be made on the fish used for studying relative coordination. For example, if one takes a fin whose central rhythm exerts a strong, dominant influence upon

another fin rhythm as long as it (the former) is active, and the first fin is passively oscillated during a period of spontaneous arrest, *no* periodic effects of relative coordination appear in the 'dependent' rhythm. Thus, mere excitation of receptors through passive movement of the muscles does not exert a coordination effect.

5. On the other hand, in some cases it has proved possible – through transection of all nerves and extirpation of the entire fin muscula- ture operated by the dominant rhythm – to demonstrate con- tinued operation of the central process underlying the fin's action. The dependent rhythm, which spontaneously reappeared soon after the dissection, at first emerged in a uniform manner and then – under the same conditions as those which had previously produced a transition from absolute to relative coordination – gave way to a corresponding periodic form, just as before. This behaviour can presumably only be explained through continued operation of the dominant central rhythm, despite destruction of the entire peripheral structure associated with it.

6. Finally, the association between the respiratory and the locomotor rhythms discussed in detail below (p. 114 et seq.), which can lead to the emergence of relative and absolute coordination between the two, clearly indicates the existence of a close intrinsic relation- ship between them. Since the automatic nature of the respiratory rhythm can be regarded as fully established even for fish (Adrian, 1932), this relationship further supports a comparable evaluation of the locomotor rhythm as an automatic process.

The experiments outlined above exclude any possible assumption that there is a reflex origin for the rhythmic process and that 'reflex' coordination is present. Instead, the entire process is fundamentally capable of taking place even when all specific afferent excitatory inputs are excluded. Gray and Sand (1936a, b) came to a parallel conclusion following similar experiments on fish of the shark family. Of course, this conclusion does *not* affect the fact that *stimuli* of all kinds can, on the other hand, *exert effects* upon the central process, producing elicitation, inhibition and/or modification.

The concept of locomotor automatic activity agrees with the earlier experimental results obtained by Graham Brown (1916) from dogs, with results reported by various authors – particularly Paul Weiss (1936) – working with amphibians, and finally with recent results from the recordings of oscillations in electrical potential taken even in completely isolated sections of a nervous system. In some clear-cut cases, these electrical discharges are arranged in series, which quite evidently occur in the tempo of locomotor movement (for example, see the experiments conducted by Schriever and von Holst on the ventral nerve cord of the earthworm; an explanatory illustration is given in von Holst, 1937d).

FUNCTIONAL DIVISION BETWEEN AUTOMATIC AND MOTOR CENTRAL ELEMENTS

A number of observations and experiments present considerable difficulties for the hypothesis that the same motor elements which transmit impulses to the muscles are also the source of the automatic rhythm. The situation at once becomes clear, however, if one follows the likely assumption that there are two types of element in the spinal cord – one type which produces the central rhythmic process, and the second type (the actual motor cell), which is alternately stimulated and inhibited by the first and which itself transmits the muscle impulses. The data in question are assembled below.

1. In general, one observes in cases of relative coordination between two rhythms that the amplitude of the fin beat with the dominant rhythm parallels the intensity of its influence upon the dependent rhythm, i.e. the extent of periodicity in the latter. However, this general rule does show certain characteristic exceptions; Fig. 2.41 provides an example. Here, a brief pricking stimulus has caused the pectoral fin to produce one beat of double amplitude, followed by one of quadruple amplitude. In line with this, one should expect a considerable resultant alteration in the observed periodicity of the dependent rhythm at this point. However, the 'triplet' periodicity occurring at this point (based upon combined operation of superimposition and the M-effect) is *de facto* identical with that occurring previously. From this, one can conclude that nothing can have changed in the reciprocal relationship between the two rhythms. But, in view of the assumption that increased beat intensity of the pectoral fins must be based upon a corresponding increase in the automatic excitatory process, this is difficult to interpret. On the other hand, no difficulty is encountered if one assumes that the reflex stimulus has temporarily produced

a sharp increase in the excitability of the responding motor elements, without affecting the automatic process itself, so that the frequency of the muscle impulses is multiplied.

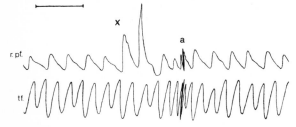

Figure 2.41. Sargus: right pectoral fin and tail fin. At x, there was a brief needle jab on the dorsal surface near the pectoral fin, which produced two intensified fin beats. The periodicity exhibited by the lower trace remains unchanged. (At a, the recording surface was halted; the irregularities in frequency just before a and just afterwards are a result of slower movement of the paper.)

2. A process which is, in principle, analogous is illustrated in Fig. 2.42. In Fig. 2.42a one can see a dominant rhythm and a dependent rhythm exhibiting superimposition by the first and proceeding with a large amplitude. In Fig. 2.42b, a weak but constant inhibitory stimulus has reduced the amplitude of the lower rhythm to about a third of its previous extent, whilst the amplitude of the upper rhythm has remained the same. If this reduction in amplitude were based upon a diminution in the underlying automatic rhythm, then one should expect a considerable increase in the superimposition effect exerted by the dominant rhythm, since the interference image is a product of the magnitude *relationship* between the two rhythms.

Yet the time curves for Fig. 2.42a, b (which are presumably

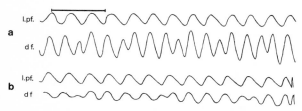

Figure 2.42. Labrus: the beating of the dorsal fin is reduced by about two-thirds in intensity in *b* by uniform pressure on the tail fin.

self-evident in this case) have an identical appearance, though their amplitudes represent an indicator of this influence. Contrary to expectation, the time curves for Fig. 2.42b are not steeper. Thus, the relationship between the two central rhythms – despite the change in amplitude of the lower rhythm – must have remained the same. This fact, too, can be understood with difficulty only if one does not assume that the inhibitory stimulus in this case has merely lowered the excitability of the motor elements responding to the dependent central rhythm, without penetrating to the automatic process itself.

3. A further phenomenon which belongs in this category is the behaviour of the upper and lower halves of the tail fin when activity resumes after an operation or following strongly inhibitory stimuli. Here, one often finds that, between the individual fin beats of constant full amplitude separated by pauses of varying lengths, there are scattered individual beats of small and extremely variable magnitude ('sub-maximal beats'). Fig. 2.43 provides two examples of this; the small beats are marked with an x.

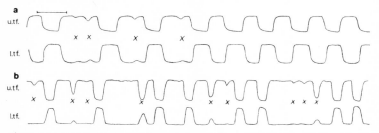

Figure 2.43. Sargus: upper and lower halves of the tail fin (u.tf. and l.tf.); only the tail is active. *a:* gradual, spontaneous appearance of activity following prior narcosis (urethane); *b:* appearance of activity following a post-operative lapse. Each point x marks one of several 'submaximal' beats of varying amplitude. (Both traces from the same fish; for explanation see text.)

In Fig. 2.43a, the small beats are in rough agreement in the two traces, whilst in Fig. 2.43b they are consistently smaller (or even undetectable) in the lower trace. If one now takes the entire recording sequence represented by the sample trace in Fig. 2.43b, measuring all the amplitudes in sequence and arranging them according to magnitude such that the amplitudes of the upper half of the tail fin produce a line which slopes upward and finally becomes horizontal (when the full, 'maximal' action is attained), then the corresponding amplitudes of the other half of the fin also fall on an upward-tending line (Fig. 2.44). However, the upward swing begins later (at 'a' in Fig. 2.44) and reaches a constant level some time after the first line (at 'c'). Thus, before 'a' only the upper half of the tail fin is active – performing sub-maximal actions – and between 'b' and 'c' the upper half is maximally operative, whilst the lower half is still performing sub-maximal actions.

This behaviour is in complete formal correspondence with that of two tissues of differing excitability which are activated by

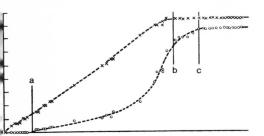

Figure 2.44. Evaluation of the amplitudes of the recording sequence represented by the sample trace in Fig. 2.43b. The lengths of the individual amplitudes of the upper half of the tail fin (x) are arranged in sequence such that they form a straight, upwards-inclined line, which changes to a horizontal line at b when the full ('maximal') amplitude is reached. The corresponding amplitudes of the lower half of the tail fin (o) then lie on a line located beneath the first. The scale on the left indicates arbitrary length units. (See text for details.)

mutually corresponding stimuli of variable intensity. Below a certain stimulus intensity, the less excitable tissue does not respond at all; at an intermediate stimulus intensity its response is always weaker than that of the other tissue, and the response maximum is only reached at a comparatively high stimulus intensity. In line with this, one could simply interpret the behaviour of the two halves of the fin, by analogy, as an outcome of the fact that the motor elements of the two have different levels of excitability, and that the impulse to each of them is at any one instant derived from the same source. This source would be an automatic process which, at any given instant, exhibits the same intensity wherever it operates. (There is another possible interpretation – that one part of the automatism itself produces two actions at any one time, one which is always more intense, and the other which occurs in parallel, but is always to some extent weaker. This is a possible conclusion; but it is far less plausible.)

4. A further argument in the same direction is derived from observation of differences in behaviour of the rhythms when action is spontaneously initiated. As mentioned at the beginning (p. 45), either the oscillation begins with a small amplitude which slowly increases, or it appears with full ('maximal') amplitude from the very outset. Transitional cases between the two can also occur. Observation of rhythms which appear in full operation from the outset shows, in fact, that emergence is always preceded by a particular, very slow movement, through which the fin is brought from the indifferent resting position to a specific starting position.

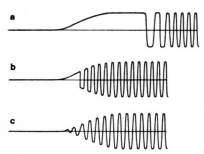

Figure 2.45. Schemata representing the various kinds of behaviour accompanying spontaneous onset of fin oscillation. (See text for explanation.)

After the onset of oscillation, this position proves to represent the location of one of the two reversal points. The schema in Fig. 2.45a demonstrates the behaviour indicated by numerous traces. However, the same fins can also start operating with a smaller amplitude, and in such cases there is never any sign of an initial transference to a starting position (Fig. 2.45c). Finally, oscillation can begin at any point during the slow, introductory movement; though in such cases one can exactly follow the initiated upward trend of the line by linking up the successive inflection points of the oscillation, as is schematically demonstrated in Fig. 2.45b.

From these observations, one has the impression that two different factors should be separable in such situations, one determining the frequency, and the other prescribing the amplitude. The two factors would seem to be extensively independent of one another. If one assumes such dualism in the automatic and the motor functions, this behaviour can easily be explained: the automatic process determines the frequency, whilst the number of motor cells excited by the process at any one time defines – other things being equal – the amplitude of the oscillation. If the two elements begin to operate together, then one should obtain traces of the type shown in Fig. 2.45c: the amplitude of oscillation gradually increases with the growing excitability of the motor cells. If, on the other hand, the automatic process begins to operate later, whilst the motor elements are already fully excitable, the activity should start at once with complete oscillations, as in Fig. 2.45a. Accordingly, we can interpret the all-or-nothing tendency described previously as a property of the automatic process, which is in many cases merely overlain by the motor cell activity and thus rendered invisible.

5. The fact that the fin has already been brought to the 'correct' starting position, when the automatism exhibits delayed emergence, is possibly correlated with the fact that the automatic process (as is shown by many traces) can apparently only come to a halt, as a general rule, following completion of a full action, i.e. that the fin (when the motor elements are still operative) can only come to a halt at one end of the reversal points, and not at some arbitrary position, as long as there are no additional incidental stimuli (cf. for example, Figs 2.43b and 2.54).

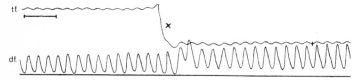

Figure 2.46. Labrus: tail fin and dorsal fin – pectoral fins out of action. The upper rhythm exhibits active, restricted 'accompanying movements' paralleling the lower rhythm. At x, there is a spontaneous, isolated action of the upper rhythm, which alters the activity level of the lower rhythm. (For a further example of individual action, see Fig. 2.45a.)

This interpretation is also in agreement with the fact that when one rhythm which is superimposed upon another comes to a halt, it maintains the dependent rhythm at a constant activity level which depends upon the reversal point at which the dominant rhythm is arrested. Fig. 2.46 provides an example. In this case, relative coordination of *movement* has become coordination of *posture*.

The interpretation outlined above would also explain the behaviour described at the beginning (Fig. 2.13), where one pectoral fin rhythm reappeared with full, medium-sized or small oscillations according to the duration of prior CO_2-narcosis. According to this interpretation, in Fig. 2.13a the motor cells would have been the first to recover completely from the inhibition; in b, they would have begun to be accessible to excitation somewhat earlier than the automatic cells; and in c, the two types of cells would have become operative at the same time (or perhaps with the motor cells lagging behind).

6. Finally, assumption of a dualistic relationship between automatic and motor functions would provide an easy explanation for the frequently observed fact that the same influence can often have opposing effects on the amplitude and frequency of a rhythmic action. Many examples of this could be provided not only for the fish rhythms discussed here but also for other locomotor and respiratory movements. If one were to assume that the frequency of a specific, central process determines the frequency of a movement, and that the intensity of the same process dictates the amplitude of movement, then it would be rather difficult to give a plausible explanation for contrasting alteration of the frequency and the amplitude through a specific influence. If, on the other hand, there are additional motor elements connected between the central process and the action, and if these elements can respond in a manner different from that of the automatic-rhythmic elements, then this phenomenon is by no means difficult to understand.

Among the arguments listed for a functional separation between the automatic-rhythmic function and the motor function, perhaps not one is fully conclusive. However, the combination of these arguments renders this interpretation an extremely probable one. In addition, there is the practical point that this assumption makes it much easier to describe many phenomena – including some quite outside the special field of spinal cord physiology.

Peiper (1938) used this interpretation as a basis for describing the relationships which he had studied between the respiratory and suckling rhythms in suckling infants (see p. 123 et seq.). However, instead of speaking of automatic-rhythmic and motor elements (cells), he refers to rhythmic and motor 'centres'. This is, in fact, short and simple, but it is probably incorrect to the extent that his experiments, along with those described here, provide no indications of separate localization (in the gross sense of the word 'centre'). It is quite possible that the two elements are intimately intertwined.[1]

The applicability of this conclusion can be illustrated with one further example. Derwort (1938) recently pointed out that it is difficult to outline figures of different shapes with the hand so that all of the distances are covered at the same speed. Instead it is a general rule that when, for example, one is asked to outline first a small circle and then a large circle, the movement for the first circle is absolutely slower than that for the second, such that both circles are traced without any great difference in timing. The experimental subject is not particularly conscious of this fact. Derwort associates this 'rule of constant figure-time' with certain interpretations from Gestalt theory, according to which there is an 'attachment of form to temporal sequence as a temporal-spatial Gestalt-making process'. He regards this as involving 'a process in which time must be regarded as having a Gestalt-making function', and he interprets his observations as a 'contribution to object-constancy'. One can, however, easily discard these hypotheses if one makes the simple assumption that two kinds of central process are involved in the movement. In analogy with the automatic elements of the fish spinal cord, one process can be regarded as determining the particular direction and the time of directional change (i.e. the frequency, if the movement is repeated), and the other (i.e. based on the motor elements) defines the amplitude of the action through the number of elements and their frequency of discharge. On the basis of this assumption, the observed movements can be described as a result of the fact that the voluntary impulse has, in each case, exclusively or predominantly affected the amplitude (i.e. via the motor cells). This is in complete agreement with the general result from influencing fish rhythms (and with various other experiences with rhythmic reflexes), according to which the frequency of an action is far less open to variation than is the amplitude, and that such variation has

---

[1] In order to avoid misunderstandings, it should be pointed out that, for example, the distinction between different (medullary and spinal) respiratory centres does not parallel the separatist principle discussed here; it overlaps. Medullary areas can quite feasibly incorporate motor elements, and spinal areas can also include automatic ganglion elements.

narrower limits. It would therefore appear to be unnecessary to introduce a far-flung hypothesis.

The relationships between different fin rays of the same fin (coordination within an 'automatism')

More detailed formulation of the central organizing process whose general operation has already been described requires more detailed information about the organization within a central rhythmic 'automatism'. Such information can be obtained by simultaneous, mechanically independent recording of the movement of two or more rays of the same fin. In the extensive dorsal fin (Fig. 2.9), each fin ray is associated with one segment, from which it receives its motor innervation. With the tail fin, the central innervation area for the fin rays (usually fourteen in number) is concentrated in about three to four segments, and with the pectoral fin the number of segments is presumably even smaller. This is indicated both by the origin of the motor nerve roots and by approximation from experiments involving multiple transection of the spinal cord and observation of the motor rhythms which can still be elicited in each case.

The following rules outlined for the relationship between rays of the same fin can be listed in an extremely summarized form since they agree in all the essential points with those listed above for the relationship between different rhythms. Within an area of associated elements, the same events are seen in microcosm as those which occur on a large scale between widely separated rhythms.

RELATIONSHIP BETWEEN THE AMPLITUDES

1. The amplitudes of oscillation of two rays of the same fin generally exhibit a certain reciprocal relationship. However, with alteration in the conditions of activity, this relationship can vary within wide limits. In the extreme case, one fin ray is completely at rest whilst the other is operating vigorously. Fig. 2.47 provides examples of this. This behaviour corresponds to that of different fin rhythms (e.g. cf. Fig. 2.12).
2. Generally inhibitory stimuli will tend to have a greater effect upon the more weakly beating fin ray, as is indicated by the

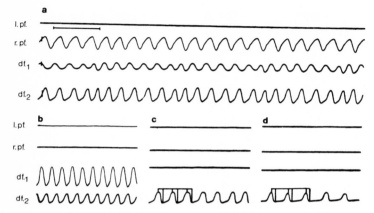

Figure 2.47. Labrus: both pectoral fins and two different dorsal fin rays separated by several segments. Specimen traces from the same fish, showing how variable the degree of activity of rays of the same fin can be under changing conditions (in this case, mild alteration in O_2 supply).

example in Fig. 2.48. This response pattern is also analogous to that shown by separate rhythms under comparable conditions.

3. One remarkable special phenomenon which is occasionally observed is abrupt (often alternating) spontaneous switching between two amplitudes, each of which is rigorously maintained. Fig. 2.49 provides an excellent example, which at the same time demonstrates that this variation in amplitude can be restricted to one fin ray (or possibly a small number of fin rays), whilst another ray of the same fin possesses an approximately constant amplitude. This observation indicates that there is periodic activa-

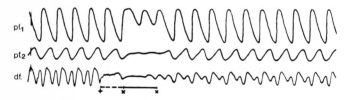

Figure 2.48. Labrus: uppermost and lowest fin ray of the same pectoral fin, and the dorsal fin. At + there is a weak pressure stimulus on the tail, and from x to x strong pressure, which inhibits the three actions to differing extents and with varying after-effects.

tion and inactivation of discrete, small automatic cell groups (in an all-or-nothing manner) and that these groups are associated only with a restricted motor cell region.[1]

FREQUENCY RELATIONSHIPS

1. In the great majority of cases, fin rays of one and the same fin usually exhibit corresponding frequencies. However, exceptions occur (a) under very noxious conditions, such as CO_2-narcosis and strychnine poisoning (see Fig. 2.15d) and (b) when the rhythm concerned is operating only weakly, e.g. after an operative intervention or following strongly inhibitory stimuli. In some such cases, it was observed that the two fin rays had completely independent frequencies. In the remaining cases, traces were obtained which agreed with expectations based on a magnet effect or superimposition between relatively independent component rhythms. Fig. 2.50 provides an example in which one fin ray initially has a frequency three times that of the other, though the latter already exhibits conspicuous superimposition upon the first. The coactive position of the two fin rays which can be deduced from such cases of relative coordination always corresponds to their mutal phase relationship under conditions of absolute coordination.

2. When there is spontaneous activation and inactivation of a fin following the all-or-nothing principle, the fin rays are so disposed that their coactive position is assured from the outset. This relationship is analogous to that between different individual fin oscillations when activated or inactivated (cf. Fig. 2.25).

The observations cited clearly show that an entire automatic cell region is divisible into a number of subgroups, corresponding to the number of associated fin rays. These subgroups are relatively autonomous, and their relationships are evidently regulated in exactly the same fashion as those between different automatisms. Over and above this, it seems that each of these subgroups can be further divided into even smaller groupings (see upper curve in Fig. 2.49).

---

[1] Any possible interpretation based on response variations in the motor elements would seem to be excluded, since these abrupt changes in amplitude are so rigidly attached to the locomotor tempo.

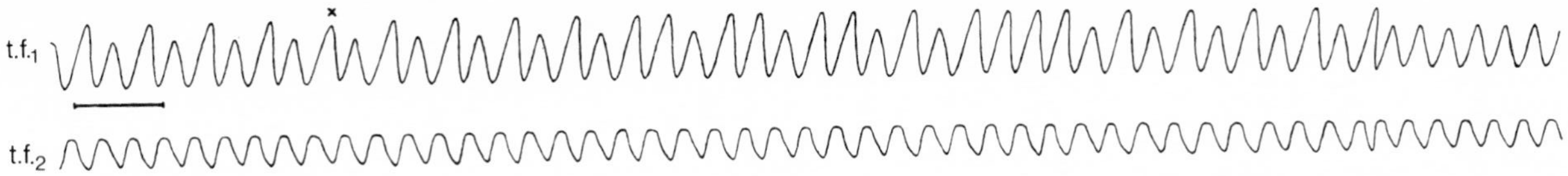

t.f.₁

t.f.₂

Figure 2.49. Corvina: one fin ray from the upper region of the tail fin (t.f.₁) and one from the central area (t.f.₂) (the rest of the fin is out of action). There is spontaneous alternation between two rigorously maintained amplitudes in the upper rhythm. (At x, there is an exceptional case of a beat which is too short.)

pf

df₁

df₂

Figure 2.50. Labrus: pectoral fin and two dorsal fin rays. The dorsal fin rhythm was inhibited roughly 10 seconds before the beginning of the trace. The two dorsal fin rays subsequently began to operate with different frequencies. At x, where the lower rhythm is momentarily accelerated as a result of relative coordination with the upper rhythm, the rapid middle rhythm is already temporarily attached to the lower one, and at the right end of the trace this attachment is almost complete.

The Behavioural Physiology of Animals and Man

PRINCIPLES GOVERNING THE PHASE-RELATIONSHIP
BETWEEN RAYS OF ONE FIN

The rays of a fin never oscillate in complete agreement (synergic-ally); one fin ray is always a fraction of a complete oscillation in advance of its neighbour when moving. (Conversely, it can equally well be said that the first is almost a complete oscillation behind the latter.) As a result, the movement of the entire fin is 'wave-like'. The 'waves' pass from the front to the back along the dorsal fin; over the tail fin and the pectoral fin they pass downwards. (Certain stimuli will sometimes produce temporary directional reversal of the wave in the tail fin.) The temporal gap with which the movement of one fin ray follows that of another (= the time which passes before the second fin ray reaches the same point of inflection as that just passed by the first: briefly, the *temporal distance*) is only rigorously constant when there is a constant oscillation frequency and when constant physiological conditions prevail. Otherwise, the following regular departures are observed.

1. The temporal distance increases and decreases in proportion to the action time (i.e. in inverse proportion to the frequency) fol-lowing any alteration in frequency, whatever its cause (through stimuli, alteration in the water supply, change of temperature, nar-cosis, strychnine application, etc.). This rule can also be expressed in the converse sense: the *spatial phase relationship* between the fin rays remains *the same, independent of the frequency at any given instant*. This is expressed schematically in Fig. 2.51. From this figure, one can see the following: If the temporal distance between the rays were always constant, the spatial phase relation-ship between the fin rays – and thus the momentary length of the waves – would necessarily change with any alteration in fre-quency (cf. Fig. 2.51a, b). If, on the other hand, the spatial phase relationship remains the same, any change in frequency is only accompanied by a change in speed of wave propagation (cf. Fig. 2.51a, c). If the first case were realized, it would be reasonable to assume that coordination of the fin rays is associated with an excitatory process of variable speed passing from one component rhythm to another. Since *de facto* it is the second case which is realized, such an assumption becomes obsolete. It has already emerged previously that the governing principle (even within a

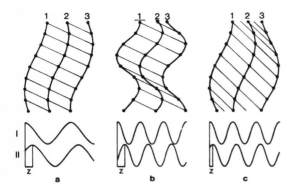

Figure 2.51. Schemata illustrating the reciprocal relationships between the individual rays of an oscillating fin. The three lines drawn in the upper parts of *a, b* and *c* (1, 2, 3) represent the wave-like fin profile at three instants of the movement, whilst the nine points lying on each of the lines represent nine fin rays. The pairs of sinus curves (I and II) in the lower parts of *a, b* and *c* represent the graphic registration of two of these fin rays (the first and the third). z is the *temporal distance* between the movements of these two fin rays. The schema *b* represents a case where the temporal distances between the fin rays remains constant as the frequency increases. The lines 1, 2 and 3 and the lines linking rays with the same phase show that – in comparison to *a* – the speed of wave propagation remains the same, whilst the wavelength decreases. *c* represents a case where the temporal distance is decreased in inverse proportion to the increase in frequency. In this case, the wave velocity increases and the wavelength remains the same as in *a*.

rhythm) is likely to be the M-effect, whose property – as we have seen above – is that of maintaining specific *spatial*, rather than *temporal*, relationships between different actions. This would explain the rule which has just been deduced.

2. Further regular relationships involve the action time and the temporal distance between two fin rays in all those cases where the fin concerned is in a state of relative coordination and therefore shows periodic variation in frequency. This relationship can be represented in its simplest form if time-tables are plotted for both the first and the second fin rays and then, for each point where a time value is plotted in the first table, the corresponding temporal distance value is plotted in a third table (beneath) – a *temporal distance table*. Fig. 2.52 provides a typical example.

One can see in the temporal distance table that two curves are produced. They are flatter than the time curves of the first table but they are otherwise parallel. Thus, there are similar periodic changes in the time values and the temporal distance values. From a large quantity of recorded traces evaluated in this manner, it proved possible to formulate the following rules. (1) The intensity relationship between the periodic variations in time and temporal distance with rhythms operating under relative coordination can exhibit extensive differences in different cases. At one extreme (a), the two parameters vary proportionally to the same extent; at the other (rarely achieved) extreme (b), the temporal distances remain approximately constant.

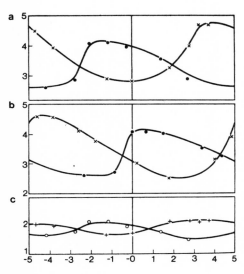

Figure 2.52. a and *b* are time-tables for two rays (separated by several segments) of a dorsal fin of *Labrus*, operating under relative coordination; *c* is a table of the temporal distances in movement of these two fin rays (explanation in text). The values for the phases and counter-phases are represented in the time-tables by ● and x, and in the temporal distance table by o and +. In contrast to Figs 2.22 and 2.34, the values at the left margin of the tables do not represent percentage departures from the prevailing mean value; instead, they represent specific time values drived from the oscillating system (viz. one-tenth of the duration of a complete dominant rhythm).

On average, the relationship between the percentage variations in time and temporal distance values is 2 : 1 with the dorsal fin and 5 : 4 with the tail fin. (2) The stated variations in time and temporal distance values always occur in the same sense (as in Fig. 2.52).

This observation agrees only *qualitatively* with the above-mentioned rule of proportionality between the time and temporal distance values with alteration of the frequency. *Quantitatively*, in cases of relative coordination the temporal distance is altered one-fifth less than would be expected from the rule with the tail fin, and half less with the dorsal fin. This phenomenon is understandable if one bears in mind the fact that the dominant influence acting upon the dorsal fin and the tail fin rhythms, which originates from the pectoral rhythm, is propagated along the spinal cord with a specific, finite speed. This influence thus acts upon the component rhythms of the dorsal fin movement (which are each one segment further removed) with a much slower sequential action than with the more closely packed component rhythms of the tail fin movement. From this point of view, the result ceases to be contradictory if one assumes a mean propagation speed of 30–40 cm/s for the dominant influence.

The derivation of time, speed and temporal distance tables for two or more rays of a given fin provides all the data necessary for exact reconstruction of the movement of all the fin rays for any form of periodicity. Fig. 2.53 provides such a reconstruction for a simple form of periodicity ('*Alternans*'), based on the assumption that the amplitudes remain constant and that there is an equally powerful influence on each of the individual component rhythms. Schema I shows the movements of the fin rays in a case where there is proportionality between the time and temporal distance values (roughly corresponding to the behaviour of the tail fin rays). Schema III demonstrates the other (rare) extreme case of constant temporal distances, and Schema II represents the intermediate situation, where the variation in temporal distance (in analogy with the behaviour of the dorsal fin rays) amounts to half of the variation in time values. One can see from a comparison of the curves that there is identical movement of all rays in succession only in Schema III, whilst in Schema I only the movements of rays 1, 9 and 17 (in parallel with rays 2 and 10, with 3 and 11, etc.) exhibit corresponding movements, which also coincide temporally, whilst rays 5 and 13 move in a mirror-image fashion. In Schema II, rays 1 and 17 exhibit corresponding movements, but 17 is delayed with respect to 1. The mirror image of these two is the curve for ray 9, between 1 and 17.[1]

The schemata in Fig. 2.53 readily demonstrate the significance of the propagation speed of the dominant influence (in this case from the top to the bottom). The image in Schema I would occur when the speed is, practically speaking, infinite; and that in Schema III

[1] The fact that neighbouring fin rays in fact *do not perform similar movements in sequence* is itself a reliable argument (should one be necessary) against the generally accepted theory that one member (in this case a fin ray), by virtue of its movement, reflexly elicits the same movement in its neighbour. In this case, how do these regular differences arise?

would appear when this speed coincides with that of wave propagation over the fin. Schema II would emerge when the former speed is about twice that of the latter.

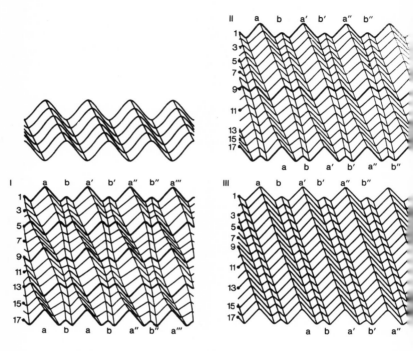

Figure 2.53. Schemata representing the movements of all rays (1–17) of a periodically oscillating fin with a simple form of periodicity ('*Alternans*' = frequency relationship of 1:2). The six lines (top left) illustrate the movement of the dominant rhythm; the sections I–III show the activity of the dependent rhythm. Schema I represents a case where the dominant influence simultaneously affects all the individual component automatisms of the dependent rhythm. Schema III shows a case where the temporal differences in effect coincide with the temporal distances in movement of the individual fin rays. Schema II represents the intermediate between the two extreme cases, I and III. (Connecting lines from a above to a below, from a' to a', etc., would in each case pass through regions of similar influence exerted by the dominant rhythm.) In order to clarify the difference between the curves in I, II and III, the sequentially occurring peaks of the 17 curves are connected with lines. In III, these are straight lines, in II they are somewhat bowed, and in I their course is extremely sinuous.

THE TEMPORAL RELATIONSHIP BETWEEN PHASE AND COUNTER-PHASE WITH THE SAME RAY AND WITH DIFFERENT RAYS

The duration of oscillations in one direction and that of oscillations in the other direction are only balanced in a very general way. With the majority of rhythms, one finds that the temporal relationship between the phase and the counter-phase can vary within wide limits, either spontaneously or as a result of certain influences. This applies just as much to symmetrical rhythms (dorsal and tail fin) as to asymmetrical ones (pectoral fin). An example of spontaneous alteration of this relationship is given in Fig. 2.47a–d; the extreme is represented by Fig. 2.47d. Comparison of different fin rays in this respect leads to identification of a number of regularities in behaviour:

1a. The temporal relationship between the phase and the counter-phase can behave in a fundamentally different manner with two rays of a uniformly beating fin; but such extensive differences are usually only transient.

1b. Statistical evaluation of a large number of curves shows that alterations in the temporal relationship between phases and counter-phases of different fin rays predominantly occur in parallel.

2a. Even in cases of relative coordination, the magnitudes of periodic oscillation in the phase and counter-phase times can frequently exhibit extensive differences.

2b. In this case, too, the relationship changes occur in parallel with different fin rays (when statistically evaluated).

2c. If the phase and counter-phase already exhibit differences in duration with uniform, independent oscillation, then with the addition of a magnet effect the slower phase generally exhibits a relatively greater variation in frequency (even in percentage terms) than the faster one. In other words, the 'weaker' phase of action undergoes a greater magnet effect (or superimposition effect).

3. If one compares first the relationship between the durations of the phase and the counter-phase of one fin ray under all possible conditions, and then the phase duration of one fin ray with the duration of the corresponding (i.e. similarly directed) phase of another ray under the same overall conditions, the following

result is obtained. The relationship between the corresponding phase durations of different fin rays separated by several segments is considerably more intimate than that between the phases and counter-phases of one and the same fin ray.

These rules, which are presented in an extremely summarized form, are supported by a complex statistical technique applied to the experimental material. Actual numerical data are provided in a previous publication (von Holst, 1939b), and it would be superfluous to go into the evaluation process in more detail in the present article.

DISTINCTION BETWEEN THE ELEMENTS OF THE PHASE AND COUNTER-PHASE

The regularities outlined in the preceding section show concordantly that the central automatic component process which determines the oscillating movement in one direction must be regarded as fundamentally separate from its partner, which determines the movement in the opposite direction, since both can operate with extremely variable conditions of activity and susceptibility to stimuli, magnet effects and the like. On the other hand, there is a close correlation between the component processes producing movement in the same direction in neighbouring segments, despite the fact that these processes do not operate synergically, but with a certain phase lag.

It is reasonable to conclude from this fact that there must be *distinct automatic elements which produce the phases and counter-phases of a given movement.* Accordingly, a component automatism should be pictured as consisting of two alternately operating subgroups of automatic cells. As is dictated by the very fact that these operate alternately, proportionality between the phase duration and the temporal distance, which is described above (p. 62 et seq.; p. 97) as typical for any coordination through an M-effect, also governs these two cell groups. It is therefore quite reasonable to conclude that the M-effect is the factor which ensures their coordination. Consequently, the problem of 'reciprocal innervation' – i.e. the fact that antagonistic muscles are generally innervated in an alternating fashion – would be incorporated as a special case in a general theory of coordination, which can explain in a uniform manner all possible phase relationships (including synergy and antagonism).

Relative coordination as a phenomenon

The following data support the validity of this interpretation:

1. The phenomenon of rhythmic antagonism (i.e. of reciprocal innervation) is coupled (not only in fish, but quite generally) with the phenomenon of maintenance of specific, *different* phase relationships. In all extremities which perform roughly ellipsoid or otherwise complex movements (e.g. the legs in running movements of man and animals; the wings in bird flight, etc.), several antagonistic pairs of muscles are operative at the same time. They are neither synergic nor antagonistic, but innervated with specific, different phase distances (in analogy with the muscles of neighbouring fin rays). Every special theory of reciprocal innervation necessarily failed to incorporate this phenomenon, whilst the M-effect can explain both equally well.

2. It is characteristic of the M-effect that it is more or less eliminated by strychnine; even adjacent rays of a fin can subsequently beat with extensive independence. Accordingly, the same elimination effect should be expected for reciprocal innervation as well. In actual fact, this was described by Sherrington some time ago (1892) for the frog and was later described for mammals as well.

3. A human being is able, through 'voluntary control', to eliminate the magnet effect between different extremities (see p. 122 below). It is well known that the same applies to reciprocal innervation in humans, as was shown particularly by Bethe and Kast (1922).

The proposed interpretation – that different elements produce the phase and counter-phase of a movement – would also explain a phenomenon which occurs quite frequently with rhythms which are slowly assuming activity, and which is represented by an example in Fig. 2.54. In this case, the phase and the counter-phase are initially separated by long time intervals, which become shorter and shorter and finally disappear. One can see quite clearly that the lengths of these intervals are not the same at the two points of inflection; there is always a difference between the upper and lower points. Measurement of these intervals shows that a continuous connection can indeed be established between those on one side and those on the other, but that the two curves which are obtained have extensively independent courses.[1] This phenomenon can also be easily understood if it is assumed that two different automatic cell groups are responsible for the phase and the counter-phase, and that

[1] An example curve is given elsewhere (von Holst, 1936a, Fig. 16).

in this case – as a result of differing degrees of suppression – each requires its own recovery time before resuming operation.

The factual content of this section can be cautiously formulated in the following way. It is extremely probable that there are different central elements for a movement and its counter-movement, and it is at least consistently conceivable (certainly not improbable) that coordination of these elements is effected by a reciprocal magnet effect.

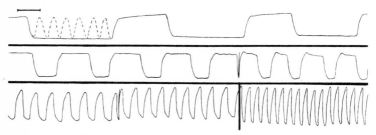

Figure 2.54. Sargus: one fin ray of the upper half of the tail fin. Samples from a long recording with slow resumption of the tail fin rhythm (interrupted by intervals of gradually decreasing length) following a post-operative pause. There is an interval of about 8 minutes between the first and the last traces. At the top left, an interval has been filled out with dotted oscillations to indicate the action frequency of the rhythm (in the absence of pauses).

Precise formulation of the theory of relative coordination

The previous section provides all the necessary data to justify an attempt at more exact visualization of the central coordination process. The theory can be briefly expressed as follows. Two kinds of ganglion element are involved in the locomotor process – the motor elements and the automatic-rhythmic elements. The former determine (*ceteris paribus*) through their number and action frequency no more than the amplitude of movement, whilst the latter decide the frequency (and similarly affect the amplitude according to their number). The automatic elements are aggregated in small groups, which can themselves be further subdivided in some instances (Fig. 2.49). The individual elements in a group operate synergically. The groups are combined in pairs which operate antagonistically (in alternation) and thus produce the alternating phases and counter-phases of the central activity. Neighbouring pairs operate with certain small phase distances. A large number of such paired groups is combined to give a relatively cohesive unit ('automatism') which supplies the musculature of a fin. There exist between different units of this kind in the spinal cord certain phase relationships, analogous to those existing between the groups within an automatism, which vary according to the component elements of two automatisms compared in any given case. All of the relationships listed are based upon reciprocal magnet effects between the groups. The magnet effect is opposed by the stabilization tendency of the automatic groups, which – according to their prevailing levels of activity – tend to operate at quite different inherent frequencies. Within an automatism which is free from any influence, the reciprocal M-effect is usually predominant: all elements operate in absolute coordination (in coaction). Between different automatisms, the stabilization tendency of the individual rhythms is frequently predominant (with our experimental subjects): the frequencies remain different, but the magnet effect makes itself noticeable through the production of characteristic forms of periodicity. If all the elements of the affected rhythm retain their operating tempo, even if this is periodically modified by the foreign influence, then a 'pure' magnet effect is present. Another possibility is that some component elements of each group are entirely subjected to the M-effect of the other rhythm, whilst the others remain independent (just as these elements can behave separately in other ways – see Fig. 2.49). In such a case, one automatism produces in the other an 'outlyer' of varying size which operates in coaction. Thus, the affected automatism falls into two components of differing frequency, which are superimposed upon one another in their effect upon the motor cells at any instant. This is 'pure' superimposition. If, in a situation of this kind, the independent component is additionally exposed to a magnet influence (pure M-effect) – as is often the case – there is an overall combination of superimposition and magnet effect.

This extremely simple suggested interpretation explains all of the details described without any supplementary assumptions. In addition, it traces the superimposition phenomenon back to the magnet effect and thus explains the close relationship between the two which was discussed on p. 80. The division of a group of otherwise synergically operating elements into two subgroups has the result that when there is addition of the two processes (with superimposition) the maximum always lies where the two groups happen to be operating in the same sense – and this relationship of course coincides with the coactive position.

This interpretation also explains all of the complex forms of periodicity to be explained below, which can arise in the cooperation of more than two different rhythms. All such forms of periodicity can be calculated in advance from this interpretation and the formal description of the M-effect.

Unanswered questions

The theory sketched out above leaves a number of unanswered questions.

1. So far, no statement can be made about the physico-chemical nature of the automatic-rhythmic process.

2. The physical nature of the magnet effect, which is presumably closely correlated with that of the central rhythm, remains unexplained. So far, the concept of the M-effect has only been derived from the phenomenon of regular modification of the affected rhythm, which represents no more than the end result of the central interplay of the forces which can be deduced.

3. One further factor which has simply been accepted is the fact that there are different coactive positions towards which the M-effect operates. The question as to why a certain phase relationship is aimed for with two particular fin rays, whilst a different relationship is aimed for with two others, would remain unanswered even in the face of complete understanding of the nature of the M-effect. One might be tempted to relate this to certain constitutional properties within the ganglion mass; but the following – so far omitted – fact speaks against this: the coactive position is *not* absolutely constant. Such constancy is only present over short periods of time. If one compares the phase relationships which are aimed for in successive traces which are separated by a number of hours or even days, they are found to differ greatly under certain conditions. Therefore, the wavelength of the fin oscillation is, similarly, not maintained constant; it gradually changes with alteration of the phase relationship between neighbouring fin rays.[1] This phenomenon presents a considerable

[1] In an intact animal, this change could presumably occur quite abruptly, with the participation of the brain.

obstacle to understanding of the physiology of central coordination since it shows that the activity equilibrium aimed for at any particular instant by the automatic groups cannot itself be based upon structural data, since it varies with the physiological condition of the components. Thus, the small, rapidly changing central equilibrium states which represent the subject of this article are embedded in large and relatively more stable ones (at least in the case of 'medulla-operated fish').

4. Finally, in correlation with point 3, it is difficult to understand how it is possible that an automatic cell group is able to exert a range of specifically distinct influences upon various others (with which it possesses coactive positions which *differ* in each case). Let us take as an example the simple case of two rhythms which exert mutual effects and which each contain only three automatic subgroups (i.e. three fin rays), a, b and c as opposed to a', b' and c'. The position would be relatively simple if there were effects exerted only by a upon b, by b upon c, by a upon a', by b upon b' and by c upon c'. In fact, however, the situation is such that each of the six elements exerts an effect upon all of the others (i.e. has a 'quintuple' influence). This fact would seem to be important since most relationships within the central ganglion mass may be based upon such *complex reticular influences*. In order to understand this, one could draw upon the well-known experiments performed by P. Weiss (1936). He has produced a number of compelling arguments to demonstrate that – according to the new formulation of his *resonance theory* – the individual ganglia respond in any given instance to a *selection* of excitatory processes, rather than to an arbitrary range. Following this theory, we can assume in our example that there is *not* transmission of *five 'different'* effects in five directions from each group pair, but that *all elements concord* in transmitting in all directions a periodic sequence of 'excitations', to which each automatic group responds in its own (functionally, but not structurally, specific) manner, differing from that of the others. Thus, with a process operating uniformly through the entire system, each component element adopts its own peculiar rôle. In view of our present scarce information in this field, this interpretation seems to be perhaps the most plausible. (For further details, see von Holst, 1939b.)

Survey of special forms of relative coordination in fish and comparison with other coordination phenomena

The previous selection of phenomena for discussion was aimed at a demonstration of the fundamental principles. This section is intended as a survey of the widely varying special expressions of these principles, grouped according to their possible parallels with other coordination phenomena.

SIMPLE FREQUENCY RELATIONSHIPS: COMPARISON WITH HEART COORDINATION

The types of simple periodicity which can arise between two automatisms, according to the proposed theory, when one is dominant and the other is dependent, can be arranged in a schema in such a way that one can also deduce the possible transitional forms which can exist between them. A schema constructed in this way is given in Fig. 2.55. If, on the other hand, the actually occurring periodic forms of fish rhythms are arranged according to their empirically determined transitional possibilities, an exactly corresponding picture is obtained.

The representation in Fig. 2.55 is, for reasons of space, restricted to cases in which the frequency difference of the two rhythms has a relationship no greater than 1:2 and is no less than 4:5 (aside from absolute coordination). The schema is set out as follows. The continuous, thick-line rhythm (marked 'domin') is dominant over all the others, whilst all the other curves are dependent upon this one rhythm. The periodic forms which arise in each case depend firstly upon the frequency of the dependent rhythm (which is represented as a dotted trace for each observed case in lines 4 and 5, in the absence of any influence by the dominant rhythm) and secondly upon the nature of the influence itself. Three possibilities are distinguished for the latter: pure magnet effect of the dominant upon the dependent rhythm (lines 1 and 8); pure superimposition (i.e. separation of a small component group, which follows the dominant rhythm; lines 3 and 6); and, finally, combined superimposition and magnet effect (lines 2 and 7). Over and above this, the lines I–IV represent the behaviour of a slow rhythm of the all-or-nothing type.

The columns B–E contain the forms of periodicity which arise under these conditions, whilst columns A and G show only absolute coordination in order to demonstrate possible transitions to absolute coordination. Column F contains frequency relationships which are not new with respect to E (in lines 1–4, F and E with respect to

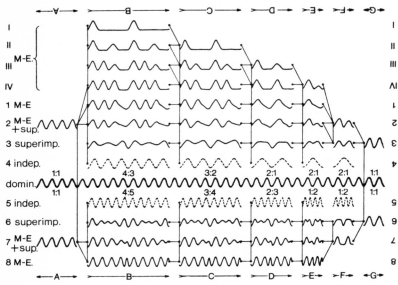

Figure 2.55. Schema for the derivation of simple forms of periodicity of one locomotor oscillation influenced by another, uniform, rhythm. Transitional possibilities can also be deduced. The schema can be read in the normal position or after rotation through 180° (and then from left to right), since the traces of fish rhythms (according to the operation of the recording lever) present themselves in two mirror-image forms. (See text for further details.)

D); but it shows enhanced influence from the dominant rhythm, and thus transitional stages to the absolute coordination in column G. The connecting lines passing in different directions between the points marked at the beginnings or the ends of the traces are intended as indicators for forms of periodicity which are closely related to one another and can interchange abruptly or smoothly.

It is probably best to provide a number of concrete examples in order to explain how the schema should be used. Let us assume, for example, that we have a dependent rhythm of an all-or-nothing type which has been arrested by an inhibitory stimulus and which is resuming activity. It at first produces isolated actions (IB), which gradually occur at shorter intervals (IIC) until a 2 : 1 relationship is achieved (IVD). From then onwards, there is gradual transition through a 3 : 2 relationship (IVC) and a 4 : 3 relationship (IVB) to final achievement of absolute coordination (A). This developmental sequence is not the only possible one. If there is a completely weakened rhythm, which also has a tendency to produce frequent incomplete actions, the sequence is as follows: IB → IIC → IIID → IIIC → IIIB → IVB → A. If the tendency for interval formation is slight, the sequence can also be: IB → IIC → IIID → IVD → IVE → 2F → G.

To take another example: let us assume that there are two rhythms in a state of absolute coordination and that one rhythm (the dependent one) is gradually slowed down. If the entire automatism becomes separated from the dominant one, the sequence goes from A → 1B → 1C → 1D. This leads to the 2 : 1 relationship, which in most cases represents an extreme from which a change can only occur in reverse or in a different direction. (These transitions from A to 1B, from 1B to 1C and so on are abrupt when there is a strong magnet effect; otherwise, there are smooth transitional forms which are not contained in the schema.) A fundamentally corresponding sequence occurs when the rhythm is accelerated: A → 8B → 8C → 8D → 8E; again arriving at a 1 : 2 relationship. If, in these two cases, a component group of the dependent rhythm remains fixed in the tempo of the dependent one, the series does not follow lines 1 or 8, but rather lines 2 or 7. If, however, one component group remains completely dependent upon the dominant rhythm, whilst the other becomes faster or slower without being influenced, then periodicity of the pure superimposition type occurs, as shown in lines 4 and 6.

The sequences listed (1, 2, 3 and 6, 7, 8) all lead from absolute coordination (A) to a 1 : 2 relationship, passing first through long forms of periodicity and then through increasingly shorter forms; but there is another direct possibility for such a transition, which is contained in the columns E–G. There is a possibility of abrupt transfer of the entire dependent rhythm (or a component of varying size) from a 1 : 1 to a 1 : 2 relationship (or vice versa). If a very small part of the dependent rhythm separates and adopts an accelerated frequency, becoming maintained by the dominant rhythm in the stable equilibrium position of 1 : 2, the periodic form shown in 7F is achieved, with an external appearance fairly close to that of a

1 : 1 rhythm. With increase in number of the separated elements, the curve adopts the form shown in 7E; and when all the cells have joined the rapid group, the form 8E is assumed. Conversely, when a small part changes to half frequency, the form 2F is produced; as the component group grows, 2E and finally 2D appear. In this case, therefore, the change from a 1 : 1 to a 1 : 2 or 2 : 1 relationship is achieved initially through strong superimposition, then weak superimposition and finally with a pure magnet effect.

Among this collection of simple forms of periodicity, some types deserve special attention since they exhibit extensive formal correspondence with those which have long been recognized for coordination among the components of the vertebrate heart (and to some extent of the coelenterate medusa as well). These formal parallels emerge where: (1) the rhythms concerned are of the all-or-nothing type, which is analogous to the 'all-or-nothing' response of the heart,[1] and (2) the dominant rhythm is the more rapid, which corresponds to the general rule that the more rapid component in the heart is always the 'leader'. If the individual fish rhythms are compared to the pulsating sections of the heart, the analogies can be summarized as follows:

1. A spontaneously emerging individual rhythm frequently has a slower frequency initially, and its amplitude may sometimes vary somewhat at first: the first beat is markedly oversized, the second is too small, the third is a little too big, and so on (see Fig. 2.13a). Corresponding effects are known for the heart.

2. As long as the dependent rhythm is not quite able to follow the leading one, it performs (according to its ability to follow) one or a specific number of actions in concordance and then omits an action (see Fig. 2.19a–c). In the first case, a frequency relationship of 1 : 2 is produced; in the second, the so-called 'Wenckebach periods' occur. These are characterized – both in the heart and in the fish rhythm – by the fact that the dependent rhythm lags more and more behind the leading rhythm before it is completely omitted (Fig. 2.19c).

3. An *'Alternans'* form can occur with heart and fish rhythms under

---

[1] Previously, I have referred more or less to an 'all-or-nothing' tendency of the automatism in the spinal cord but this is not really appropriate, since this concept is based on a relationship between stimulus intensity and extent of the response. With fish rhythms, on the other hand, 'spontaneous' actions are involved and one cannot say anything about the intensity of any existing primary central 'stimulus'.

conditions which are partially analogous. The *'Alternans'* (amplitudes a) can in both cases assume either a form where one large beat (relative to the normal action) is followed by one which is correspondingly too small, etc. (Figs. 2.38, 2.39, 2.48), or a form where one beat has a normal amplitude whilst the second is too small (Fig. 2.49, left). In the latter case, both the fish rhythms and the heart often exhibit 'partial asystole'. The fish rhythm seems to lack an analogue of the 'total hypodynamy' which occurs in the heart and the medusa (Bethe, 1937b).

4. In some further forms of periodicity which will not be dealt with here (e.g. Lucianic periodicity, frequency *Alternans*, bigemini and so on), coordinated components of the heart and fish rhythms show extensive formal analogies.

The existence of these parallels between heart and spinal cord coordination is important to the extent that it shows that even systems with such distinct functional plans can nevertheless exhibit formal analogies. But it is a general rule that one can only conclude with great caution from this latter fact that the same factors are operative.[1] The coordinating factor in the heart is an excitatory process which passes at intervals, whilst in the spinal cord there is a continuous influence whose efficacity can be demonstrated at any moment in time.

In the heart, this process is only capable of excitation; with the spinal cord its mode of operation can vary according to the situation in which it encounters the dependent rhythm.

A fundamental and obvious *difference* between the two systems – even under analogous conditions – is the response to 'supplementary stimuli' which are well known with the heart, where they generally leave the basic rhythm unaffected (as long as they do not affect the source of stimulus production itself) and are followed by a compensatory interval. Of course, the application of analogous 'supplementary stimuli' to the spinal cord preparation through peripheral stimulation is impossible, since the latter produces manifold changes in the entire action position. Instead, one has to rely on

---

[1] I myself initially believed, on identifying these analogies, that one could assume what was obviously too extensive a functional parallelism between the two systems (von Holst, 1935a). Only detailed study of relative coordination uncovered this error. Bethe (1937a) later operated in a similar manner (which I regard as erroneous) to extend conclusions regarding the coordination of the heart and the medusa to the central coordination process in fish (cf. also von Holst, 1938a).

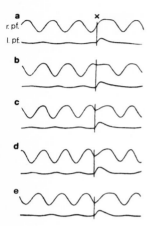

Figure 2.56. Serranus: right and left pectoral fins. The right fin is active, whilst the left performs slight 'accompanying movements'. *a–e:* sections form a long recording, in which there is sporadic occurrence of single, spontaneous 'supplementary impulses' (marked by x). Since the gill operculum exhibited simultaneous twitches, the impulses were assumed to originate from the medulla. Each impulse provokes both fins to perform simultaneous forward beats. There are no 'compensatory intervals'.

chance occurrence of spontaneous, specific impulses from other central regions. A series of such cases is given in Fig. 2.56, where sporadic impulses from the medulla caused two pectoral fins to perform forward beats in unison. One of the fins was not active in this case, and its beat can be taken as an indicator for the 'supplementary stimulus'. It can be seen that, whatever the moment of appearance of this supplementary stimulus, the supplementary action is never followed by a compensatory interval. This indicates that, as is entirely to be expected from the prevailing situation, the source of the

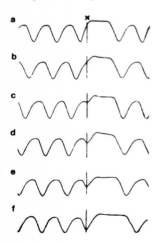

Figure 2.57. Serranus: right pectoral fin – the same preparation as in Fig. 2.56 (the left pectoral fin rhythm is omitted here). *a–f:* a series of extracts with responses to spontaneous 'supplementary impulses' of the kind seen in Fig. 2.56. In this case the beat is more pronounced (the curve rises beyond its normal level) and it is followed by an interval. This interval is, however, not 'compensatory' in the sense of unaffected continuation of the previous operating rhythm.

113

stimulus itself (the automatism) is affected every time by the supplementary impulse.

Even in cases where an evidently extremely powerful 'supplementary stimulus' of this kind elicits an inordinately great supplementary action, which can then be followed by an interval (which is differently situated as compared with the analogous case of the heart or the medusa), there is no close relationship between the length of this interval and the tempo of the rhythm, as is shown by a series of traces in Fig. 2.57.

LOCOMOTION AND RESPIRATION

The relationship between locomotor and respiratory rhythms cannot be studied in most 'medulla-operated fish' since respiration disappears following the operation. However, in a number of cases incomplete, infrequent respiratory movements (of the gill opercula) reappeared later, and clear-cut relationships between these movements and the fin rhythms were observed. Examples are already available from Figs. 2.56 and 2.57, where the 'respiratory' impulse acted as a 'supplemen-

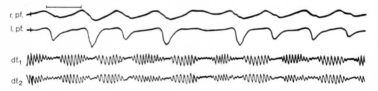

Figure 2.58. Labrus: both pectoral fins and two rays of the dorsal fin recorded during a strychnine experiment (cf. Fig. 2.15). Paradoxical relationship between the oscillations occurs: irregular, synergic pectoral fin beats (rhythmically coincident with slight movements of the gill opercula) are accompanied by simultaneous inhibition of the dorsal fin action.

tary stimulus'. More paradoxical cases occurred in cases of strychnine poisoning, where the previously arrested pectoral fins each performed one beat in synchrony with one 'respiratory' impulse, whilst the actively operating dorsal fin rhythm was weakened, to emerge in full strength later on (Fig. 2.58).

Clear cases of relative coordination between respiratory and fin movements were provided by experiments on non-operated fishes of other species. For example, an intact *Trigla* (gurnard) – a substrate-living fish with large, colourful pectoral fins which are spread as a 'protective' or 'frightening' response to every optical stimulus – exhibits to a greater or lesser extent simultaneous retardation of opercular opening.[1] The relationship becomes even more pronounced under mild narcosis (urethane), which led to a large number of cases of rhythmic fin-spreading in time with the respiratory tempo, and sometimes with a frequency relationship of $1:2$ (analogous to the fin rhythms in Fig. 2.19a) over a fairly long period. Comparable observations on a correlation between respiratory and fin rhythms, which is particularly evident under narcosis, have been reported previously (von Holst, 1934c).

My colleague Le Mare (1936) has observed with sharks a phenomenon which is interesting in this context. Medulla-operated sharks exhibit responses different from those of spinally-operated sharks (where the transection is made posterior to the medulla) when the O_2 supply is cut off or when CO_2 is introduced. With sharks, both kinds of preparations produce spontaneous, rhythmic locomotor movements along the trunk (sinuous movements). However, although these rhythms are uniformly suppressed in spinally-operated sharks under the above conditions, in medulla-operated subjects the rhythms frequently adopt a periodically increasing and decreasing form. This is analogous to the periodic respiratory pattern described under the name 'Cheyne-Stokes respiration' for various animals and, occasionally, for human beings. This finding and a number of others indicate with high probability that the periodic onset and cut-off of the fin rhythms of our preparations under certain conditions is also directly based on medullary influences.

From these incomplete data it would seem that both the M-effect and other more general forms of influence can operate between respiratory and locomotor rhythms. Overall, this field has unfortunately been too little touched by comparative studies. There are a number of studies on the relationship between respiration and locomotion, e.g. in the embryonic stages of fish (Babak, 1911) and mammals (Barcroft *et al.*, 1936). Horse riders are acquainted with the synchronization of the respiratory and gallop rhythms in horses, and the common (but not regular) $1:2$ relationship between the two rhythms is well known for the trot. Analogous phenomena have been described for decerebrated dogs (Olefirenko, 1937). However,

---

[1] A sample illustration is provided in another publication (von Holst, 1937a).

there has not yet been a systematic investigation of the detailed relationships involved.

THE LOCOMOTOR RHYTHM AND THE 'DEFENCE' REFLEX

In analogy with the relationship between the respiratory rhythm and the 'frightening' reflex of the pectoral fins found in *Trigla*, *Sargus* exhibits a relationship between the locomotor rhythm and a defence response, which consists of marked spreading of the powerful, pointed anterior dorsal fin rays (which otherwise have no locomotor significance). In the intact fish, this 'defence' reflex has apparently no relationship with the locomotor rhythm; spreading of the fin rays occurs in an arhythmic fashion as an exclusive response to specific excitation (e.g. some external threat, or fighting with conspecifics, etc.). In medulla-operated fish, the same response regularly occurs as a completely uniform rhythmic phenomenon under conditions of pronounced O_2 deficiency; in fact, it sometimes emerges when all of the locomotor oscillations are at a standstill. However, if swimming rhythms, such as that of the pectoral fins, are operative, the spreading rhythm and the swimming rhythm are in many cases observed to occur synchronously for a period of time.

This phenomenon, along with the corresponding observations on *Trigla*, leads one to suspect that even under normal conditions quite arhythmic, specific 'reflexes' are similarly based upon automatic-rhythmic processes, such as the respiratory and locomotor rhythms, but that their activity is modified to a tremendous extent by stimuli and central impulses. (Consider, for example, the enormous distortion of the rhythmic medullary respiratory impulses imposed by human speaking and singing.)

CHANGES IN GAIT

In a number of cases, medulla-operated fish (*Sargus*, *Smaris* and *Serranus*) exhibited a phenomenon which one can also frequently observe with intact fish when they are induced to change their speed abruptly. Because of the correspondence with a comparable phenomenon in mammals, this will be referred to as a change in *gait*. Fig. 2.59 provides a typical example.

One can see how, at a particular point, both pectoral fins abruptly change to a different amplitude and frequency, and to a different phase relationship (i.e. from synchrony to alternation). This corresponds to 'gait-switching' (e.g. from a gallop to a trot) in horses, which is similarly associated with abrupt change in amplitude and frequency, along with alteration of the phase relationship. The third rhythm in Fig. 2.59 has, up to the moment of change, an independent, more rapid frequency; subsequently, it is forced to adopt the tempo of the pectoral rhythms, and it also exhibits an increased amplitude. This latter fact agrees with the observation that the pectoral rhythms *add* their M-effects and superimposition effects upon the dorsal fin rhythm when they are oscillating alternately and that

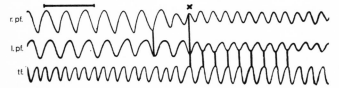

Figure 2.59. Sargus: pectoral fins and tail fin. At x there is an abrupt transference from one 'gait' to another as a result of gradual increase in the water supply from zero to the normal level.

their effects are *subtracted* (in this case mutually cancelled) when the fins are operating synchronously (cf. Fig. 2.36).

What is remarkable about this phenomenon is the fact that it can be regularly evoked experimentally by certain alterations of the water supply. In the free-moving, intact fish, on the other hand, the same gait change occurs 'spontaneously' or in response to optical and other stimuli. It is presumably evident that the elicitation in such cases does not occur through an alteration of O_2 supply to the ganglion elements concerned. From this, one can conclude that *the same changes* in central activity can arise *in quite different ways*. This important conclusion is supported by the fact that all of the described alterations in coordination and activity of the rhythms, which are elicited in medulla-operated fish by changes in water supply, narcosis, specific stimuli and the like, can occur in intact fish without any necessity for these specific treatments. Instead, these effects are produced by influences in which the brain is presumably intimately involved, and they often occur in quite rapid sequence without any visible cause.

117

COMPLEX PERIODICITY INVOLVING MORE THAN TWO
RHYTHMS: PARALLELS TO LAWS OF GESTALT PSYCHOLOGY

Finally, a short description should be given of the enormous variety
of periodic forms which can occur when more than two rhythms
of differing frequency are related to one another. This kind of interaction demonstrates a number of remarkable principles; but it represents a vast subject which can only be dealt with in summarized
form here (see von Holst, 1939a for details). In general, it has
emerged that it is possible to calculate theoretically all complex
forms of periodicity so far observed from the principles derived from
simple cases of relative coordination. Thus, these complex forms
represent nothing which is really new; instead they simply illustrate
the variety of phenomena which can arise when the number of
equivalent, interrelated components increases. Any attempt to produce a schema, analogous to that in Fig. 2.55, containing all possibilities for the relationship of three or more rhythms, is doomed to
failure.

If one considers only whole number relationships, in which the
most rapid rhythm performs no more than six actions in one period,
there are nineteen different possible numerical combinations, though
many of these (e.g. $1:2:6$, $1:5:6$) hardly ever occur, since the frequencies of the three rhythms rarely differ to this extent. A number
of other combinations – in accordance with theoretical expectations
– do not occur in practice, since they do not represent states of stable
equilibrium. Thus, the numerical relationships $1:2:4$, $1:3:4$ and
$3:4:6$ remain as stable cases, which are in fact those most frequently realized in actual preparations. However, every one of these
cases – according to the reciprocal intensity relationship of the magnet effects and to the strength of the superimposition effect imposed

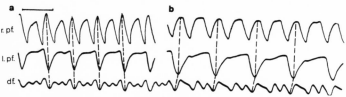

Figure 2.60. Labrus: both pectoral fins and the dorsal fin. *a* and *b* taken from
different preparations. Examples of periodicity with a frequency relationship of
$1:2:3$ (cf. also the left half of Fig. 2.8d).

on the individual rhythms – can emerge in a whole range of different forms. Fig. 2.60a and b, along with Fig. 2.8d, give three variants for the rhythm relationship 1 : 2 : 3 which show extensive formal differences.

The range of possible periodic forms varies from preparation to preparation, depending upon the success with which the frequencies and reciprocal influences can be modified by different treatments. Fig. 2.8 gives samples from ten different periodic forms of one preparation, and it is easy to derive one pattern from another.

In Fig. 2.8a and b, there is a 1 : 1 : 3 relationship, but in b the upper rhythm has already begun to exhibit a small, alternating interposed action. In d (left), this has developed to give a full '*Alternans*': the alternating action amplitude is based upon pronounced superimposition by the second rhythm (whose peaks point downwards), in that absence of the latter leaves the upper rhythm unmagnified. In 8d (centre), there is abrupt transfer to a 3 : 2 : 5 relationship. The middle rhythm from then on allows only omission of every third action of the upper one, and correspondingly only every third action of the above rhythm is unmagnified. Through analogous abrupt changes, the relationship 4 : 3 : 6 in Fig. 2.8e is achieved. Continuing in the same direction, one passes through the forms shown in Fig. 2.8h (5 : 4 : 8) and 2.8i (6 : 5 : 10) to that shown in 2.8c, where both upper rhythms finally exhibit the same tempo (1 : 1 : 2). In Fig. 2.8g, this relationship has given way to one of 2 : 2 : 3, in which the upper rhythm undergoes pronounced superimposition by the lower one, whilst the middle one is free of superimposition (3 : 2 – *Alternans*; see schema in Fig. 2.55 3C). Finally, Fig. 2.8f represents a periodic alternation of the combinations which are continually maintained in Fig. 2.8e and h.

The rules which emerge from comparative investigation of many periodic forms of this kind are the following:

1. Only a certain proportion of the extremely wide range of possible combinations is actually realized, and the forms actually observed are distinguished from the other possible forms by their greater stability.
2. This stability is expressed in the fact that, with gradual, smooth alteration of the external or internal action conditions, the periodic forms initially maintain themselves until attainment of a critical limiting situation, in which transference to another equilib-

rium relationship (i.e. a different periodic form) occurs – usually in an abrupt manner. In its turn, this new form will maintain itself within a particular range of conditions. Thus, each of these equilibrium states possesses an internal stability which renders it to a certain degree independent of change in external factors (including those *in the CNS*, which are *external* to these rhythm relationships).

3. The stability which characterizes each of these periodic forms as a whole does not apply to their individual temporal subdivisions, in which *dis*equilibrium states are more likely to occur, tending towards a *different* phase relationship between the rhythms. However, within the temporal unit of the entire period, all of these disequilibria are exactly balanced. One can therefore say that even temporally separated 'components' interact with one another in the production of the whole phenomenon ('temporal Gestalt').

4. There is a general tendency towards transference to equilibrium states of ever-increasing stability. The degree of stability increases with the simplicity of the frequency relationships. Under otherwise similar conditions, the most stable coordination form is the absolute one, i.e. that with a tempo relationship of $1:1$ ($1:1:1$ in the case of three rhythms). The next most stable is $1:2$ (viz. $1:1:2$, $1:2:2$ or $1:2:4$), then $1:3$ and $2:3$, and so on. Increasing degree of complexity is accompanied by decreasing stability.

These rules are remarkably comparable to those which have been described in Gestalt psychology for the 'Gestalts' in optical and acoustic perception[1]: (1) the tendency towards stable equilibrium states; (2) the independence and autonomous nature of these functional equilibria; (3) the fact that there is reciprocal interaction between separated components within the framework of the whole; (4) the tendency towards formation of relationships of maximum simplicity ('Law of Simplicity') – all are found in both instances. W. Koehler (1923) has shown with different physical dynamic systems that formally corresponding rules can play a role in every case. The fact that the same phenomena have now been identified in the interaction of organizational forces within the CNS is doubtless *more than an arbitrary additional physical model* for the illustration of certain rules governing 'psychological' events. However inaccessible the rela-

---

[1] For example, see Kaffka (1931); a presentation of the Gestalt problem is given by Matthaei (1929).

tionship between psychological events and physiological processes in the CNS may be to human logic, the discovery of functional parallels in both areas will necessarily play a considerable part in any future attempt at interpretation (from a scientific point of view).

Relative coordination in mammals and man

The question as to how far the interpretation developed here is generally applicable is primarily dependent upon the occurrence of the phenomenon of relative coordination in other areas. It has so far been reported for mammals and man (von Holst, 1938b), and it is probable that it also occurs occasionally in invertebrates; though presumably it does not play such an important part in any of these animals as it does in fish.

RELATIVE COORDINATION IN MAMMALS

Relative coordination has occasionally been identified in horses, and more frequently in dogs. It occurs in addition to the various gaits, emerging from walking, alternating gait and trot, and consists of an acceleration of the fore-leg rhythms relative to those of the hind-legs. Figs 2.7 and 2.6 provide a film sequence and a group of explanatory curves. Evaluation of films has produced the following data.

(1) As a rule, the fore-leg and hind-leg rhythms are periodic, such that an increase in frequency of the first is accompanied by a simultaneous decrease in frequency of the latter, and vice versa. (2) Less frequently, and only as a transitional occurrence, either the fore-leg or the hind-leg rhythms may remain uniform. In this case, however, the variation in frequency of the other (dependent) rhythms is relatively more pronounced than when both vary. (3) Even before transference to relative coordination, the leg rhythms usually exhibit relative phase displacement (slight advance of the 'faster' rhythm). Thus, even with absolute coordination, the relationships are not fixed but can vary continuously within certain limits. (4) The prevailing form of relative coordination is not noticeably modified by distraction of attention (barking, etc.), so it is almost certainly not based upon an 'intention' to move in such a manner.

The phenomena described are therefore completely analogous to those found with reciprocal or unilateral magnet effects in fish; and this justifies the assumption that the same coordination factor occurs in mammals. No indications of the existence of superimposition effects were found, but this is probably correlated with the fact that locomotion takes place on a firm substrate, which excludes, on mechanical grounds, any extensive and independent variations in amplitude of movement.

RELATIVE COORDINATION IN THE HEALTHY HUMAN BEING

In adults

In human beings, relative coordination occurs under various conditions. It emerges spontaneously in certain learned locomotor patterns, such as in the 'crawl' of swimmers, where the more rapid leg rhythm often exhibits involuntary variation in frequency and amplitude following the tempo of the slower arm rhythm. However, exact measurements of this are lacking. In addition, typical relative coordination appears in voluntary arm and leg movements when these are intended to occur at different rhythms. Two examples are given in Fig. 2.4.

On the basis of numerous recorded traces taken from unprepared experimental subjects, who were instructed to close their eyes and move two levers with each arm operating at a different, comfortable rhythm, the following conclusions were reached:

1. Many experimental subjects only succeed in moving the two arms independently for short periods (if at all), whilst some can move their arms completely independently, at least for certain periods of time. Practice plays a large part in the attainment of a maximal degree of independence and the same applies (at least initially) to concentration. Distraction of attention usually leads to rapid mutual attachment of the rhythms.

2. The influence between the two arm rhythms can be either unilateral or reciprocal, and this condition can change several times within a short period of time (Fig. 2.4 II).

3. All of the forms of periodicity produced can be matched by corresponding examples from fish rhythms. If one applies the interpretation developed for fish rhythms, the human being accordingly exhibits both the magnet effect and superimposition, usually the two together.

4. The phase relationship towards which the subject tends (coactive position) is that of alternating or synchronous arm swinging.
5. The resulting forms of periodicity are produced involuntarily by the experimental subjects and – at least initially, when the situation is entirely new – this is not consciously realized. On the other hand, subjects do consciously perceive a certain 'resistance' opposing the intention to perform independent movements.

These data clearly indicate that the interpretation developed above can be applied to human beings as well. Point 5 indicates that the locus of coordination is presumably to be sought in the lower levels of the CNS, perhaps in part within the spinal cord.

Relative coordination in the suckling infant

A. Peiper (1938) has reported a series of observations on human babies which indicate that there is relative coordination between the respiratory and suckling movements, and between respiratory movements and the periodic appearance of *singultus* ('hiccuping').

When the suckling rhythm (which is usually more rapid) commences, there is usually a struggle with the respiratory rhythm ('period of disruption'), frequently following temporary arrest of the latter. The more powerful suckling rhythm is superimposed upon the respiratory rhythm so that (for example) periodicity of the *'Alternans'* type can appear. Finally, the respiratory rhythm usually becomes attached to the suckling rhythm, adopting the same frequency (or, less frequently, a tempo relationship of 1 : 2). After cessation of the suckling movement, the respiratory centre tends to continue for a short period with the accelerated suckling tempo. This latter phenomenon, along with the cut-off of respiration at the onset of suckling, represent special cases which, as a rule, do not occur with fish rhythms (however, cf. Fig. 2.58, involving a strychnine-treated fish in which the dorsal fin rhythm is inhibited each time the pectoral fin action starts functioning). The remaining phenomena are analogous to corresponding cases among locomotor rhythms.[1]

[1] This relationship between the respiratory and suckling rhythms indicates that the 'suckling reflex' is also automatic in nature. The same interpretation applies to my own observations that, in the sleeping infant, the suckling rhythm can emerge spontaneously in the form of 'Cheyne-Stokes' periodicity (i.e. in extremely regular series of 6–10 movements each, separated by long intervals) at times when the child is known, by past experience, to be hungry.

An even more clear-cut form of relative coordination in the suckling infant can be seen between respiratory movements and *singultus.* According to Peiper, the latter can, in principle, emerge in any phase of respiration, but it does so predominantly (for certain periods of time, exclusively) in that respiratory phase which is simultaneously somewhat lengthened because of its emergence. This behaviour is completely analogous to that of two fin rhythms between which there is relative coordination, in cases where one is only active from time to time (Fig. 2.26). Statistical evaluation of the phase relationships in the infant would produce a maximum at a particular phase position (coactive position) as in the case of fin rhythms.

RELATIVE COORDINATION UNDER PATHOLOGICAL CONDITIONS

Very recently, R. Jung (1939) made the interesting observation that typical relative coordination can also appear in the tremors of people afflicted with Parkinson's disease. Among other things, Jung was able to draw the following conclusions from multiple recording traces of muscular action currents and movements of the extremities.

1. The frequency of movement usually differs between the different regions of the body; but the reciprocal innervation of the antagonistic muscles exhibits constancy.
2. Where it occurs, absolute coordination appears only in transient form. By contrast, one frequently finds forms of relative coordination which are entirely analogous to those in fish rhythms, and which can be interpreted as the product of superimposition and magnet effects within the rhythms. Fig. 2.61a and b provides examples. The two lower curves in Fig. 2.61a are independent of one another at the right and left margins; in the middle, one of the rhythms (the more rapid) exhibits pronounced periodicity (superimposition by another rhythm).

   In Fig. 2.61b, the lower, slower curve exhibits a periodic form involving both an M-effect and superimposition. (From o onwards, there is absolute coordination between the two rhythms.) Thus, both the slower and the faster rhythms can be periodic (i.e. dependent) in form.
3. Relative coordination in such cases is most pronounced between

symmetrical extremities, and there is an apparent general tendency towards alternating movement of the two (the coactive relationship is usually that of alternation in movement of the extremities).

4. The reciprocal relationships vary spontaneously and they can also

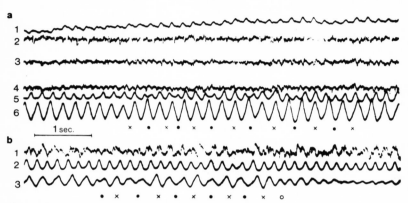

Figure 2.61. Examples of relative coordination in Parkinsonian tremors (extracts from recordings made by R. Jung, 1939). *a:* 1 = right biceps (mechanical recording); 2 = right triceps (electrical); 3 = right extensor carpi (electrical); 4 = right flexor carpi (electrical); 5 and 6 = left and right foot (mechanical recording). The maxima and the minima of the periodic variations in amplitude of the more rapid left foot movements are marked below the curves with x and ●, respectively (cf. the schema in Fig. 2.55, curve 6C).
b: 1 = flexor carpi (electrical recording); 2 and 3 = left and right foot (mechanical recording). The periodic, dependent rhythm in this case is the slower one. The maxima and minima in amplitude are marked with x and ● respectively. After o, there is transition to absolute coordination (cf. the periodicity illustrated in curve 2B in the schema of Fig. 2.55).
  In the curves for the feet, a rise in the curve signifies foot-raising. The trace extracts *a* and *b* were taken from the same patient with an interval of a few minutes; they illustrate the variations in frequency and reciprocal dependency relationships of the foot rhythms.

be influenced by impinging stimuli. Pain stimuli usually favour temporary synchronization of the rhythms, but they can also have an 'inhibitory' action.

These parallels between Parkinsonian tremors and relative coordination in medulla-operated fish are primarily important for interpretation of this pathological phenomenon. In Jung's opinion, they would support a possible phylogenetic interpretation, according to which destructive processes in higher organizational centres would conceivably favour the emergence of lower, otherwise obscured, mechanisms which are incorporated into normal movement. Secondly, Jung's findings also provide firm support for my previously expressed opinion (1938b) that the factors of relative coordination are probably to be found in the lower regions of the CNS in healthy human beings. In other words, even with voluntary movements of the arms or legs at different rhythms, the resulting forms of periodicity emerge *despite*, and *not because of*, this 'voluntary element'.

Discussion of the results

If one surveys the range of response patterns of a spontaneously active medulla-operated animal, one gains the impression that its behaviour is intermediate between that of a free-moving, intact fish and a 'reflex organism' (e.g. an animal – fish, frog or mammal – with the spinal cord transected). For it shares with the latter the rigid constancy and reproducibility of a given pattern under standardized conditions (hence its high precision, which is scarcely excelled by a physical model); whilst, in common with the former, it exhibits spontaneous activity and a wide spectrum of possibilities, in the face of which a 'reflex preparation' seems poorly and rigidly endowed.[1] Both points are of general significance. (1) The precision of occurrence found with even the most complex forms of periodicity demonstrates the invalidity of the common opinion, particularly prevalent in recent times, that central nervous coordination is fundamentally inaccessible to any exact analysis. This opinion is probably a product of the utilization of unsuitable methods (cf. von Holst, 1935g). (2) The spontaneous activity and extensive variability/plasti-

---

[1] The fact that the medulla-operated fish is, on the other hand, restricted to certain possibilities in comparison with the intact fish is demonstrated, for example, by comparison of the behaviour following abbreviation or partial amputation of the surface of a fin. In the medulla-operated fish, the movements are exactly the same after the operation as before (following, in some cases, the passage of a short period of standstill evoked by the amputation stimulus). On the other hand, the intact fish at once responds with increase in the amplitude and frequency, or – when the fin is beating in rigid, absolute coordination with another – with a slight increase in frequency of the two coordinated fins and pronounced increase in amplitude of the partially amputated fin. Thus, as in many other cases, the brain is indispensable for such regulation (even if the usually accepted conclusion that the latter is 'localized' in the brain is unproven and improbable).

city of behaviour has the automatic effect that any attempt to describe in an organized fashion the activity of medulla-operated animals – along with that of the intact organisms – solely on the basis of concepts derived from classical physiological study of the CNS in 'reflex organisms', either rapidly encounters difficulty or, so to speak, 'bypasses' an explanation.[1]

It would seem to be far more difficult to explain the wide and variable range of spontaneous activity of the medulla-operated or intact fish on the basis of concepts developed from 'reflex preparations' than to argue the other way around, passing from the pluripotent to the limited case (in a manner which I regard as more natural) and interpreting the 'reflex' activity on the basis of concepts developed from medulla-operated fish. With the 'reflex preparation' one finds a number of fixed, distinct relationships which can lead (and have led) to the establishment of 'laws'. With the medulla-operated fish, these represent (at least partially) special cases in a continuous series of variations which in the case of the spinally-operated fish is reduced to a number of fragments, since the activity level is extremely restricted. In the place of many possible equilibrium states, one is left with no more than a few, perhaps only one. It is obvious that study of such preparations leads to inappropriate fixed mechanical/structural concepts of the coordination process, rather than to purely dynamic interpretations.[2] Any attempt to 'explain', for example, the fact that a specific stimulus elicits a certain response from a spinally-operated animal by evoking the exis-

---

[1] A 'bypass' of this kind is involved, from our point of view, when (for example) the concept of the 'reflex' – which has a clear-cut content in simple cases, such as that of the endogenous reflex – is frequently extended such that it becomes synonymous with the very general term 'central nervous activity'. In the process, the word loses its character and its descriptive value, but at the same time it unjustifiably retains the appearance of a precise term.

[2] This is not intended as a dismissal of the method in other contexts, where it can be extremely valuable. These remarks simply emphasize that numerous reflex studies carried out on fish by the author together with various co-workers (von Holst, 1935–1938) have demonstrated with increasing conviction that the so-called motor spinal reflexes usually incorporate within them the entire complexity of central organization, which must be explained if one wishes to 'understand' higher level behaviour patterns. Thus, the theory that these are 'building bricks' of higher levels of behaviour does not advance us very far. Secondly, we were unable to identify any evidence for the hypothesis that 'natural' component elements of normal behaviour (rather than artificial fragments) are actually involved, as long as one does not apply the 'law of habit', according to which the concept is regarded as established from the outset.

tence of pre-established reflex pathways, loses its credibility when one observes that this response only occurs when certain additional conditions are maintained, and that under other conditions different (sometimes directly opposed) responses can occur. Structural relationships are, of course, present; but they are present in great variety. Where the 'excitation' concerned can sometimes follow one 'pathway' (definitely a completely inappropriate term) and sometimes another, this must presumably be based upon variations in the pre-existing activity level (i.e. the general dynamic equilibrium position) endogenous to the affected elements. One can interpret in the same manner the well-known fact that the intact animal can respond to a particular reflex stimulus in a quite variable, unpredictable manner. This phenomenon is necessarily extremely problematical for any structurally defined theory of coordination. On the other hand, the actual problem should, in my opinion, more justifiably be sought from a converse point of view: Why is it that, following transection of the spinal cord or under a number of other experimental conditions, certain stimuli will subsequently only elicit specific locomotor acts? In the spontaneously inactive 'reflex preparation', the wide range of possible responses to the same external influences is frequently (though indeed not always) restricted to only one response. The so-called 'Expansion Rule' of von Uexküll, according to which the same stimulus can under certain conditions produce movement in different directions in correlation with the initial position of the extremity involved, would thus be regarded as a retained special case of the general conditioned nature of most stimulus effects.

The general irrelevance of the quality of the factor eliciting a response in a spontaneously active medulla-operated fish is perhaps most evident in those cases where characteristic action patterns (e.g. changes in gait, or typical 'attack' or 'frightening' responses) are elicited in a regular fashion by influences (such as alteration in O_2 supply) which definitely do not act as releasers with the intact animal. The transition from one central activity state to another can apparently be achieved by a wide range of treatments.

Setting out from this point of view, it is only a short additional step to reach a theory of 'automaticity' of central activity and coordination, according to which any specific releasing stimuli are fundamentally quite unnecessary. This automaticity, which can be regarded as confirmed for the object of our research and has been demonstrated (or at least indicated) by recent investigations on nervous activity currents in other animals, only appears problematical from the point of view of 'reflex' coordination. The fact that at the lower levels of the CNS many organized processes can continue even in the absence of peripheral reflex stimuli, as long as the general physiological operating conditions are maintained, is in perfect agreement with findings showing continued functioning of the cerebral cortex when stimuli are excluded, during sleep and often under narcosis as well (Berger, Kornmüller, Adrian and many others). This also agrees with the recent observation that even higher, complex 'instinctive behaviour patterns', which require specific 'releasers' under normal conditions, tend to emerge spontaneously as abrupt 'endogenously activated vacuum activities' when the stimuli concerned are withheld for long periods (Lorenz, 1937b; Lorenz and Tinbergen, 1939).

These phenomena must have some relevance to the attempt to give a more detailed formulation of the reflex concept with coordinated activities. For when one attempts to determine the range of processes which can occur spontaneously (in the strictest sense) within the CNS, as has been done in this article, one is far more able to correctly evaluate the versatile role which peripheral stimuli can play in these processes. For example, when the locomotor rhythm itself continues to appear automatically (= without reflex stimuli) under suitable conditions, the fact that under different conditions a specific stimulus will evoke a fragment of a locomotor movement (a motor reflex) can presumably only mean that the stimulus in this case briefly provides the missing condition for appearance of the rhythm. If this stimulus evokes only an isolated phase of the movement, such as stretching or bending of a limb in a quadruped (cf. Fig. 2.10), then this should be regarded as the ultimate extreme of brief activation of the automatic elements. To use a physical illustration: in this case the evoked 'oscillation' would be maximally 'suppressed', whereas at the other extreme (continuous automatic rhythmicity) the central mechanism would be 'unsuppressed'. Cases of brief rhythmic processes ('rebounding' and so on) can be easily arranged in this series, which can in fact be practically demonstrated in a medulla-operated fish which is slowly recovering from the operation. Thus, in such cases the stimulus *primarily* has only a general activity-increasing effect. The fact that the movement which appears as a result must vary according to the initial central situation is self-evident under this interpretation of the role of the stimulus.

In addition to this, however, the same stimulus can simultaneously exhibit a number of other effects. In the first place, there are further primary effects such as the elicitation of 'genuine' reflex responses (e.g. of the endogenous reflex type of higher vertebrates), which are characterized by the fact that they can only be elicited by this stimulus (and not by other releasers) and cannot appear automatically. Secondly, the stimulus can produce various *secondary* and tertiary effects. A secondary effect would be represented by a situation where the elements set in motion by the stimulus exert recipro-

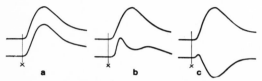

Figure 2.62. Schemata illustrating the response of the pectoral fin rhythms of a still spontaneously inactive medulla-operated fish to stimulation of the dorsal fin (at x). The stimulus itself in each case evokes forward movement of the two fins (2.62a; see also Fig. 2.10a). If a mild M-effect is exerted upon the lower rhythm by the upper one, the action of the former is modified in *b*. A strong effect produces the situation shown in *c*. (*b* and *c* are taken from various original curves.)

cal effects upon one another (magnet effect, superimposition, etc.) which further modify their action. For example, with a poorly excitable medulla-operated fish which is still responding as a 'reflex preparation', stimulation of the dorsal fin produces simultaneous forward movement of both pectoral fins (see Fig. 2.10a). However, this movement of the fins is in opposition to their central coaction, according to which they should in fact beat in alternation. Thus, if the mere stimulus to activity is supplemented by an M-effect of one rhythm upon the other, the course of the latter is secondarily modified to a greater or lesser extent, dependent upon its intensity, as is illustrated in the schema of Fig. 2.62b and c. It is far from unlikely that such central processes, which are secondary with respect to the stimulus, are incorporated in many motor responses.

Analogous secondary stimulus effects occur when the stimulus acts upon an already active prepared fish such that the intensity relationship between the individual active elements, and thus their reciprocal relationship, is modified, as shown in Figs 2.16 and 2.28. In Fig.

2.48, an analogous effect is achieved because the inhibitory stimulus after-effect is of different duration for the various similarly affected rhythms. The pectoral fin immediately returns to its previous level, whilst the dorsal fin rhythm remains subdued for some time and consequently exhibits considerably more pronounced dependence upon the pectoral rhythm than before (well-defined '*Alternans*').

As an example of *tertiary* effects of a stimulus one can take those cases where the rhythmic action reappears with an initially increased vigour following suspension of an intense inhibitory stimulus. The phenomenon (known as 'rebounding' or 'postinhibitory amplification') occurs in fish under certain conditions (Le Mare, 1936) and it was recognized some time ago in mammals. From the present investigation we have seen that, following removal of influences which *exclusively* produce marked retardation in the automatic elements (such as the M-effect of a much slower rhythm) or arrest them completely (suitable inhibitory stimuli), the elements resume their previous inherent frequencies and intensity of activity (cf. Figs 2.16 and 2.48) and never exceed these levels. We have also seen that one stimulus can exhibit simultaneously different effects of differing duration. Thus, this 'rebounding' phenomenon can be easily interpreted on the basis of the assumption that the stimulus concerned (1) directly paralyses the automatic elements, and in addition (2) produces a situation *favourable* to the activity, which persists beyond the first effect when the stimulus ceases. As a result, the automatic elements would, in a certain sense relative to the original stimulus, experience a 'tertiary' transient augmentation on resuming activity.[1]

In this direction, which has merely been touched upon, it would seem that simple and more complex responses can be separated, theoretically and experimentally, into components, and it is conceivable that one could eventually reach what is in principle a quite simple interpretation of central nervous function. Within such a theoretical framework, the endogenous activity of the CNS and its spontaneous internal organizational forces would inevitably play a major role. These are phenomena which are in many ways unrepresented in classical physiological studies of the CNS. The consequent restriction of central function to mere formation of connections

---

[1] Where the amplitude alone is increased by this 'postinhibitory reflex amplification', only the motor cells would be affected by the stimulus after effect.

between the stimulus and the motor impulse has been criticized in various ways. On the one hand, criticism has come from psychologists who are actually predominantly concerned with the features which the nervous system is not able to produce according to theories based on the reflex concept. W. Koehler (1933) has pronounced upon this subject in his confrontation with 'behaviourism':

> The idea that the interest of psychologists should be directed to the 'stimuli' on one hand and to the 'responses' on the other is a natural correlate of the concept of the organism, and particularly the nervous system, as an entity lacking any endogenous, characteristic processes of its own. Since no other concrete interpretations of the manner of functioning of the nervous system have been developed, this seductive formula[1] of behaviourism has achieved wide acceptance, and it is apparently thought that it simply expresses the application of scientific conceptualization to the general task of psychology.

In this situation, it is naturally no chance effect that the first attempt to take some account of the *spontaneous* automatic organizational processes in the CNS should, at the same time, provide a bridge from neurophysiology to at least one field of psychology (i.e. *Gestalt* psychology). This thus leads to an interpretational area which is characterized by the very fact that the vocabulary of pre-existing neurophysiology is entirely irrelevant. Our subject of research therefore assumes, in this respect as well, a mediating role which may well be of some importance.

If we now turn to some of the individual facts demonstrated by the analysis of medulla-operated fish, it becomes even more evident that simple description with the exclusive use of borrowed concepts — in particular those of 'excitation' and 'inhibition' — is not entirely successful. Even the common case interpreted as that of pure superimposition (e.g. Fig. 2.33) cannot be correctly described as 'periodic inhibition' or the like, since the 'inhibition' of the frequency is, of course, always accompanied by 'excitation' of the speed of action, and vice versa. These terms are just as inadequate for the description of phenomena interpreted here as the product of M-effects. To mention just one case, we have seen that under certain conditions an

---

[1] Namely the formula according to which the nervous system merely links the stimulus and the movement following the pattern of simple reflexes.

'inhibitory' M-effect reverses the direction of action of the influenced rhythm (Fig. 2.28, x); yet 'inhibition' could at the most produce complete arrest, and never directional reversal. Many other examples demonstrate, in a similar manner, that the terms 'excitation' and 'inhibition' are inadequate for the description of these phenomena. This is, above all, true for the wide variety of complex periodic forms appearing with more than two rhythms of differing frequency, although these present no particular difficulty in terms of our interpretation.

Conversely, when one sets out from the medulla-operated preparation, a whole range of phenomena already demonstrated for motor reflexes seem to fit into the interpretation without requiring supplementary hypotheses. The phenomenon of 'reciprocal innervation', which has already been mentioned, can be regarded as one of many possible organizational forms which, in principle, have the same origin. Brief mention should also be made of the possibility of interpreting as consequences of the M-effect many cases in which one reflex action 'makes way' for another particular action and, conversely, cases where two responses 'inhibit' or exclude one another. Without doubt, the M-effect is also intimately involved in the functional organization of individual components of an extremity which is set in motion in a reflex manner. One can also easily interpret the common 'accompanying movements' of muscle groups at various distances from one another as cases of superimposition. Mention of many other possible applications of our interpretation can be omitted; its validity must be tested experimentally. It can be assumed that, following this direction, one would be able to separate off many cases of apparent 'excitation', 'facilitation', and the like, and that these concepts themselves could be more exactly formulated in their application to the residue of observations.

This entire discussion is intended as no more than a signpost in a new direction, which cannot appropriately be developed in detail here. It is my conviction that future interpretations will have to move in this kind of direction if one is concerned with constructing a comprehensive, uncontradictory conceptual framework, unless the present factual evidence is to be left untouched, or, conversely, the many important experimental results of classical reflex physiology are to be pushed aside. A much clearer separation of fact and theory will consequently be seen as necessary in many cases, before it is possible to give new interpretations of experiments which are so heavy with hypothesis and have been described from the point of view of certain theories, using their terms. The author is well aware that it is difficult to avoid mistakes in this direction, and that such are contained in this paper as well, despite the attempt to demonstrate where pure observation ceases and where reflection begins.

If, in retrospect, we regard the medulla-prepared fish as a simple *model* for the mechanism of higher central organizational processes, we can infer an important fact which is useful for countering a certain pessimism in this area, which is derived from the recognition of the unlimited versatility of central nervous processes. The medulla preparation shows us that it is fundamentally possible to trace a wide range of phenomena back to the interaction of very few, well-defined factors. Accordingly, the complexity of central processes is based not upon the number of qualitatively different forces but upon the wide variability in interaction, i.e. upon the rich variety of possible dynamic equilibria.

Appendix

The diagram on p. 135 illustrates *coupling*. The individual rhythm is provided by the rotation within a vessel containing oil of a plate driven by a weight passing over a pulley. The size of the weight and the frictional resistance of the plate determine the endogenous frequency. Both rhythms are coupled with an elastic link (which only responds to pulling). If this coupling link is powerful, or the two endogenous frequencies are little different from one another, the rhythms are propagated synchronously with intermediate frequency. This is referred to as *absolute coordination*. If the coupling link is too weak, the two rhythms separate from one another, and characteristic *frequency-modulated periodicity* occurs; this is *relative coordination*. As the more rapid rhythm overtakes the slower one, it is retarded and accelerated in alternation, according to its relative phase position, whilst the slower rhythm itself is alternately accelerated and retarded. The maximum of this effect is located where one rhythm is a quarter of a cycle in front of or behind the other; the minimum of the reciprocal influence is located at the phase relationship in which the coupling link tends to freeze the rhythms – the 'requisite position' or *coactive position*.

We can also lend to one of the two rhythms in the model (e.g. the slower one) a greater *'stabilization capacity'* by attaching a heavy weight and simultaneously introducing a greater frictional resistance. This rhythm then becomes relatively immune to influence; it becomes the leading, *dominant,* rhythm.

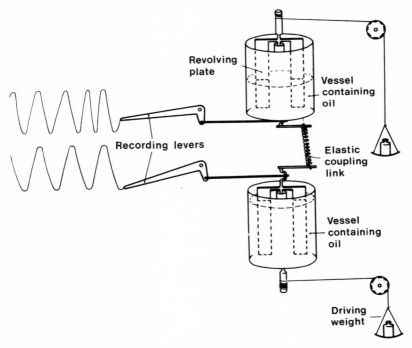

Figure 2.63. Mechanical model of central coupling of two motor rhythms with the character of a 'magnet effect'. (From von Holst, 1955b.)

Part Two
Functional structures
in the central nervous system

3 The reafference principle
(Interaction between the central nervous system and the periphery)
WITH HORST MITTELSTAEDT (1950)

Introduction

Ever since the physiology of the central nervous system (CNS) came into existence, one question has remained the focus: What regular relationship exists between the impulses which arrive in the CNS, following evocation by stimuli, and those which are then transmitted (directly or indirectly) to the periphery, i.e. between *afference* and *efference*. The CNS is portrayed as the image of an automat which produces, in a 'reflex' manner, a given travel voucher in exchange for a specific coin. With simple protective reflexes – such as sneezing and retraction following a pain stimulus – this interpretation is an appropriate one, and it has similarly been used to explain more complex responses, for example equilibrating and orienting movements. Rhythmic locomotion can also be understood on this basis, if it is assumed that every individual movement reflexly sets in motion its antagonistic partner and that every extremity provokes movement in its temporal successor (reflex-chain theory). Finally, the higher, experientially modified behaviour patterns are fitted into the picture as 'conditioned' reflexes.

This *classical reflex theory* by and large dominates the field although many facts have been recognized which do not concord with it. We know that the respiratory centre continues to operate even without rhythmic stimulus impulses, that the central locomotor rhythms of some invertebrates (von Holst, 1932, 1933, 1938b) persist without afference, and that in fish (von Holst, 1935f, Lissman, 1946) and amphibians (Weiss, 1941; Gray and Lissmann, 1946; Gray, 1950) an

139

almost negligible residue of afferent nerves suffices for continued movement of *all* parts in a coordinated fashion. In addition, analysis of *relative coordination* (von Holst, 1935–43) in arthropods, fish, mammals and men has demonstrated the existence of central organizational forces – coupling and superimposition phenomena – whose interaction leads to rules which are formally quite similar to those discovered for the subconscious organization of sensory perception in Gestalt psychology (von Holst, 1939a, 1948).

These new results resist any description using reflex terminology, and it is therefore comprehensible that, whilst they have had a certain influence upon comparative behavioural research (Lorenz, Tinbergen and others) and upon human psychology (Metzger), they have not been assimilated in studies of the actual physiology of the CNS. Even quite recent textbooks are still entirely constructed on the classical reflex concept.[1] The fact that the intact CNS is an actively operative structure in which organized processes are continuously taking place without stimulus impulses, and that even resting and sleeping represent no more than special forms of central nervous activity, strikes many physiologists as being an unscientific concept. It is believed that the only possible 'cause' of any central process must be 'the stimulus'.[2]

This attitude is, after all, understandable, for nobody will readily give up a simple theory – particularly when it is regarded as a 'fact' because of its long history – before a better one is available. The new theory must incorporate both the old *and* the new results and permit predictions above and beyond the area which one has so far been able to understand. New experiments have led us to an interpretation which we think lives up to this expectation, within demonstrable limits. This interpretation will be explained with examples, and its applicability to well-known, but previously unexplained, phenomena will be examined. The characteristic feature of this new

---

[1] This applies, for example, to the comprehensive work of Fulton (1943), *Physiology of the Nervous System*, which leads the reader from the simple spinal reflexes up to the operation of the entire nervous system (the conditioned reflex) without attributing any part to spontaneous endogenous activity or to autonomous organizational forces in the CNS.

[2] This misunderstanding probably has psychological motives as well. It is much more satisfying in view of the naïve requirement for causal explanation to be able to relate a visible motor activity of the body to a visible alteration in the environment, than to invoke invisible energy turnover within the CNS. The latter is apparently felt to be semi-psychological.

conceptual framework is a rotation of the point of attack through 180°. Rather than asking about the relationship between a given afference and the evoked efference (i.e. about the reflex), we set out *in the opposite direction* from the efference, asking: What happens in the CNS with the afference (referred to as the 'reafference') which is evoked through the effectors and receptors by the efference?

An introductory example

Let us begin with an example. If a cylinder with vertical black and white stripes is rotated around an immobile insect (e.g. the fly *Eristalis*) the animal rotates itself in the same direction – it 'attempts to stabilize its visual field' (Fig. 3.1a). This well-known 'optomotor reflex' can be immediately elicited at any time. However, as soon as the animal sets itself in motion, one can see that it performs arbitrary turning movements both in the (immobilized) striped cylinder and in an optically unstructured environment. The question then arises as to why the 'optomotor reflex' does not return the animal to its former position whenever a turning movement is initiated, since the displacement of the image of the retina is, after all, the same as with the rotating cylinder. The reflex theory answers as follows: Because the optomotor reflexes are 'inhibited' during 'spontaneous' locomotion. But this answer is erroneous!

The animal has a thin, mobile neck, and the head can be rotated through 180° around the longitudinal axis and then glued to the thorax so that the two eyes are spatially interchanged (Fig. 3.1b: see also Mittelstaedt, 1949). In this way, a situation is produced where rotation of the striped cylinder in a clockwise direction produces the same migration of the image across the retina as with anti-clockwise rotation of the cylinder around the normal animal. Accordingly, the resting animal promptly responds to rotation of the cylinder to the right with a movement of its own to the left. If the animal begins to run around in the immobilized cylinder, locomotion should occur unaffected as in the intact animal, assuming that the 'optomotor reflexes' are now inhibited. *No normal running* occurs in the striped cylinder: *Eristalis* continuously turns in small circles to the left or the right, or there are brief, pronounced right–left turning movements until the animal becomes immobile 'as if frozen' in an atypical posture. When the head is rotated back to its former position, the behaviour also returns to normal.

This result contradicts the 'reflex-inhibition' hypothesis, since it shows that, in 'spontaneous' running as well as in the rotating cylinder situation, retinal displacement exerts an effect upon locomotion. For the time being, this can be formulated as follows. The running animal 'expects' a quite specific retinal image displacement, which is neutralized when it occurs. If, on the other hand, retinal image displacement occurs in a direction *opposite* to that expected following interchange of the two eyes, this immediately elicits optomotor body-turning. But this turning movement itself magnifies the unexpected retinal image displacement, and the process is therefore self-reinforcing. As soon as any turning movement is initiated, the animal is optically propelled around in the same direction of rota-

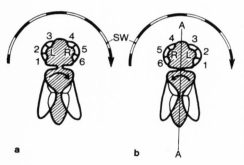

Figure 3.1. Behaviour of an insect (*Eristalis*) with a striped surface (SW) which is rotated around to the right. *a*, normal animal; *b*, following rotation of the head through 180° around the axis A–A. R, right eye; L, left eye. (The ommatidia are numbered.) The arrow drawn on the animal indicates its direction of active rotation.

tion. If it attempts a movement in the opposite sense, the same dilemma appears. The result is evidently a central catastrophe!

If this provisional interpretation is valid, we must ask: How does the CNS 'know' which particular image displacement it should expect when the animal is moving? Such knowledge could have two sources. Either the CNS preserves for a period of time certain data (regarding the efference transmitted to the legs) which are computed against the retinal image migration; or, if it does not possess this simpler capacity, it relies upon the reafference from the receptors in the moving legs to 'calculate' the direction and speed of running, in order to compute the result against the retinal reafference. These two possibilities are not mutually exclusive. We shall leave these alternatives open here and turn instead to another example which allows clearer appreciation of the situation.

More detailed formulation of the problem

In the labyrinth system of the vertebrates, there is on either side of the head a flat stone – the utriculus statolith – which is situated horizontally on a sensory surface in the normal head position. This organ responds to gravity. Experiments on fish (reported in von Holst, 1950a and c) show that a force parallel to the resting surface represents the adequate stimulus. When the head is inclined away from the normal position, the gravitational force increases sinusoidally and produces in the postural centre a similarly sinusoidal increasing disequilibrium in activity (a 'central turning tendency') which evokes motor responses which return the animal to the normal posture. This apparatus operates with great precision without habituating. In the terminology of the reflex theory, the organism is maintained exactly in its normal position through its 'postural reflexes'. However, one can often observe with all animals and in human beings that they can adopt a posture *different* from the normal position for varying periods of time. For example, fish can position themselves with an almost vertical upward or downward inclination, turn on their sides and so on when searching for food, hunting prey, fighting or taking part in courtship.

In view of the continuously evokable 'postural reflexes', how are these deviant postures possible? As before, the reflex theory would state, 'through complete or partial inhibition of the "equilibrium reflexes"'. Yet it can easily be shown that this interpretation is *not* valid. These 'requisite postures', which deviate from the norm, are actually restored through the *same* correcting movements, when disturbed by some external impulse, as those which are seen in restoration of the normal posture!

One might therefore think that the reflexes are not extinguished but simply diverted into other pathways by a higher-level switching mechanism, i.e. that the higher centre merely sets the points which determine the pathway from afference to efference (the 'steered reflexes' of W. R. Hess). This interpretation also involves an experimentally testable corollary: someone who is changing points performs the same work whatever the number of trains passing over the line. In other words, the effect of such reflex steering should not be affected by the magnitude of the afference. Once again, this is *not* the case.

One can increase the mechanical force which the statolith exerts upon the sensory surface by applying a constant centrifugal field. If the weight of the statolith is doubled in this way, there is a constant doubling of the shear stimulus produced by any departure from the normal posture. If one in fact records the frequent, spontaneous postural deviations (snout pointing upwards or downwards) in a free-swimming fish, it emerges that they become increasingly restricted with increase in weight of the statolith. The *'voluntary movement'* proves to be *dependent upon the afferent return stream* which it evokes!

To take another example. Fish orient their longitudinal axis in the direction of a water current by optically 'attaching' themselves to the immobile environment. This is also largely true when the current comes from obliquely above (or below). The greater the success in pointing the snout towards the current, the less is the energy required to prevent displacement downstream (von Holst, 1950a). If one goes on to investigate the behaviour of a fish swimming freely against a constant current within a cage, it emerges that the fish has decreasing success in maintaining its longitudinal axis parallel to the current, as one approaches a 'vertical' current direction, with increasing weight of the statoliths (Fig. 3.2). The fish does, in fact, then attempt to oppose the current *dorsally*; but this is far less successful, as it tires rapidly. This difference disappears following removal of the statoliths; whatever the strength of the mechanical field, the snout is oriented directly into the current.

Thus, one can see that the higher central factor which determines the *requisite posture* deviating from the norm is not a point-setting or steering mechanism, since the reafference indirectly produced by its command has a quantitative influence upon the posture which actually appears as a result. Despite this dismissal of the concept of the 'steered reflex', the following pronunciations on the approach to the problem and the conceptual framework involved are quite closely related to the works of W. R. Hess (although the methodology is quite different). The 'steering device' must perform *more work*, the *greater the number of trains* which pass along the line! How can this be interpreted?

We can obtain a quite simple picture if two reliably demonstrated physiological facts are taken as a basis:
1. The sensory cells of the labyrinth, like many (perhaps most) other

receptors, exhibit a continuous endogenous rhythm – the *continuous trace* prior to any stimulation. The shear stimulus of the statoliths merely increases or decreases the resting frequency, according to the direction of shear. This *automaticity* of the receptors has

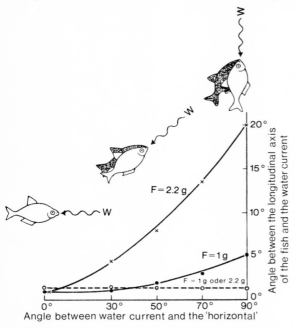

Figure 3.2. Orientation of a fish swimming freely within a cage against a constant water current. The current direction is altered from horizontal to vertical and the strength of the mechanical field (F) is increased from 1 g to 2·2 g (through a centrifugal field). (The 'vertical' is a resultant of the gravitational and centrifugal forces.) ● and x represent the intact fish; o represents the fish after an operation to remove the statoliths. The white fish outlines illustrate the behaviour of the intact fish at F = 1 g, whilst the black outlines indicate the behaviour at F = 2·2 g. W is the water current. (The current velocity is approximately one fish length per second.) The hydrostatic pressure is maintained constant, and the values are taken from five experimental series involving three fish (*Gymnocorymbus* and *Hypessobrycon*).

been verified through our own experiments on fish and also by direct electrical recording either from the afferent fibres (e.g. O. Lowenstein, 1950) or from the vestibular nucleus (Adrian, 1943).

2. A *continuous stream of impulses* links the higher and lower centres even when there is external motor inactivity. This fact has also been demonstrated for various central areas through electro-

physiological investigations, and is indirectly evident from the 'shock'-like reduction in activity which occurs on transection of certain ascending inter-central pathways. The 'spinal shock' of lower motor centres following destruction of (for example) the *tractus vestibulospinalis* corresponds to the 'shock' suffered by the left vestibular nucleus following destruction of the left labyrinth.

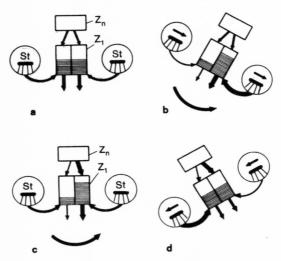

Figure 3.3. Schema explaining the interaction of higher centres (Z_n) with the lower postural centre (Z_1) and the static apparatus (St) in the postural orientation of a fish (along the longitudinal axis). The thickness of each arrow denotes the intensity of the stream of impulses (number of discharges per unit time) flowing from one component to another, and the shading of the half-centres of Z_1 indicates the momentary level of discharge (activity). St = statolith; the arrows above it in *b* and *d* indicate the direction of shear, and the large arrow in the centre of *b* and *c* indicates the direction of the rotational tendency. *a*: normal situation; *b*: rotational tendency towards the normal position following passive inclination of the animal; *c*: active (spontaneous) rotational tendency towards the left; *d*: the resulting inclined 'requisite posture'.

In both cases, a continuous activating stream of arriving impulses is disrupted.

These two preconditions are necessary for the interpretation explained in Fig. 3.3. The 'postural centre' (a complex of ganglion cells whose spatial distribution is irrelevant at this point) consists of two halves, each of which receives a stream of impulses from the statolith epithelium and from higher central areas. If both halves are charged at the same level (i.e. at the same level of activity), they both send impulse streams of equivalent intensity to the lower motor centres of the spinal cord, which in this case too have no more than a charging function (Fig. 3.3a): the well-known 'tonus'-effect of the labyrinth (Ewald; see von Holst, 1950a). If the fish is passively inclined to the right (Fig. 3.3b), the shear of the right statolith produces – as has been verified in experiments (von Holst, 1950a) – an increase in the continuous afference, whilst that of the left is reduced. The resulting difference in level between the right and the left halves of the postural centre produces an imbalanced impulse stream to the spinal cord, which brings into operation local motor apparatus which, as a whole, leads to a turning movement to the left. This process is referred to in classical terminology as the *postural reflex*.

However, a corresponding difference in level within the postural centre can also arise as a result of imbalanced impulses arriving from the higher centres (Fig. 3.3c and d). The result is *the same* motor action as before.[1] This process is known in physiology as a *'voluntary movement'*. It is instructive that the *inclined* 'requisite posture' is *protected* against passive disturbances through the afference from the statolith apparatus in exactly the same way as the previously maintained normal posture. Any passive alteration in posture will lead to a difference level within the postural centre, and thus produce a 'postural reflex' without any involvement of higher centres.

We can now consider varifiable corollaries of this model interpretation:

1. Destruction of the left statolith apparatus, as a result of simultaneous lowering of the level within the postural centre, must produce a persistent turning tendency towards the left. This tendency must exhibit a maximum when the fish is on its right side (maximal charging afference from the right) and a minimum when it is on its left side. This applies to all vertebrates and has been quantitatively validated in fish (von Holst, 1950a). The same

[1] The concept that higher centres have the function of distorting the excitatory equilibrium in lower antagonistic centres was first developed in the analysis of relative coordination in rhythmic locomotor movements (von Holst, 1936a, etc.). Recently it has been further validated by investigations of 'action currents' (Bernhard and Therman, 1947).

also applies, as is well known, to the injury or destruction of the postural centre itself (vestibular nucleus; Spiegel and Sato, 1927).

2. When the left half of the postural centre has recovered from the removal of the statolith on the same side and has been re-charged to the normal level,[1] any turning tendencies occurring with a change in posture should only exhibit half of their former value, since the left level remains constant, whilst only the right level is dependent upon posture. This has been quantitatively demonstrated in fish (von Holst, 1950a).

3. Amplification of the intensity of the mechanical field, and thus of the afference, must also amplify in a regular fashion the part played by the statolith apparatus relative to that of other equilibrating organs (e.g. the eye in fish) in postural orientation. For example, doubling of the intensity of the mechanical field exactly compensates the lack of one statolith (point 1).

4. Following exclusion of the higher centres, no further active departures from the normal posture should occur. This has also been demonstrated (Magnus, 1924, etc.).

5. Spontaneous changes in the requisite position, or those produced by other afferences passing through higher centres, should produce a *decreasing* alteration in posture with *increase* in the intensity of the mechanical field, and thus of the *shear stimulus* of the statoliths. In this point too, as we have seen, the expectation is fulfilled: the balancing of the two halves of the postural centre is achieved with approximately *the same shear*, i.e. with a correspondingly *smaller angle of inclination* of the animal when the statoliths are heavier.

6. Conversely, following bilateral exclusion of the afference, imbalance in the stream of impulses arriving from higher centres should lead to *exaggerated* movements, since in the absence of reafference – mechanically speaking – there is loss of the 'stop' which arrests the initiated movement at the right moment. This phenomenon can also be observed at any time with free-swimming fish and amphibians, and it has frequently been reported. Animals lacking the labyrinth on both sides exhibit postural alterations which are so pronounced that general tumbling often

---

[1] Our co-worker L. Schoen has quantitatively examined this central compensation process, but no further discussion of this can be conducted here.

results.[1] In terrestrial vertebrates, this behaviour is less obvious because of the loss of tonus following removal of the labyrinths and because of the pronounced part played by muscle receptors in the movement (see pp. 158–64). Nevertheless, the phenomenon can be observed following loss of just one part of the balance receptors. For example, following destruction of both horizontal semi-circular canals, a slight, unstable rocking movement of the head in the horizontal plane occurs in intention movements (particularly in birds).

Thus, one can derive a number of valid corollaries from this conceptual model – corollaries which cannot be explained by the reflex theory.

More general formulation of the reafference principle

The key feature in the example explained above is the role played by the reafference produced by active movement. This extinguishes the alteration in conditions produced in a lower centre by a motor command from a higher one, so that the former equilibrium is restored. If this afference is removed under experimental conditions,

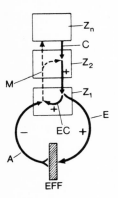

Figure 3.4. General schema for the explanation of the reafference principle. For explanation see text.

[1] Schöne (1950) has reported entirely analogous observations on insect larvae. The larva of *Dytiscus*, which normally presents its back to the light, will (for example) perform somersaults when swimming upwards head-first following blinding of the forward- and backward-directed eyes. This is apparently because there is absence of the reafferent 'stop' (stronger illumination of the forward-directed eyes and lesser illumination of the backward-directed ones) which normally arrests the correcting movement. (Schöne's own interpretation runs in a similar manner.)

if it is too large or too small, or if its sign is altered (as in the rotated head of *Eristalis*), predictable alterations in the locomotor process can occur.

It is now necessary to provide a more general formulation of this principle and then demonstrate its validity with examples from different neuromotor mechanisms.

Let us consider (Fig. 3.4) a given centre Z_1 which has sensory and motor connections with an effector EFF. This effector can be a muscle, a limb or the entire body. One or more further centres (Z_2 to Z_n) are superior to the centre Z_1. Any *command* from Z_n – i.e. a specific *change* in the stream of impulses descending to Z_1, produces an efferent series of impulses (E) from Z_1, which produces a closely correlated alteration in activity – the *efference copy*, EC – which spreads through the neighbouring ganglionic mass with a specific temporal delay. The efferent stream E flowing out to the periphery evokes the corresponding reafference A from the effector, and this reafference interacts with the efference copy. The efference and its copy can be arbitrarily marked with a plus (+), whilst the reafference is marked with a minus (−). The efference copy and the reafference exactly cancel one another out in Z_1, whilst the command descending from Z_n flows outwards without interruption. As soon as the entire afference is too large or too small, as a result of some *external influence* acting upon the *effector*, either a + or a − remains as a residue in Z_1. As we shall see, *this residue is transmitted* upwards, sometimes to the highest centres; it can be referred to as a *message*, M. The ascending message can branch in Z_2 on its way (although it *does not necessarily* do so), where once again it summates with the descending command. In this case, the system from Z_2 downwards will maintain itself in equilibrium, thus producing a *feedback system* of the kind recognized in technology.[1] Let us assume, for example, that an influence affecting the effector EFF produces an *increase* in the − afference in Z_1, in which case the ascending message to Z_2 *reduces* the + command until an equilibrium is reached once again. Conversely, an externally produced *decrease in* the − afference in Z_1 will bring about a + residue which will *amplify* the + command through Z_2. Thus, in both cases, *the efference*

[1] We owe our acquaintanceship with technological cybernetics to Dr Böhm (see also Böhm, 1950).

150

is modified until no further message is sent out by Z_1.[1] We have already encountered a feedback system of this kind in the example of postural orientation.

In addition, we can refer to the alteration of the afference which is *not* a direct consequence of an efference but arises through *external* influences (through proprioceptors or exteroceptors) as an *exafference*. The *exafference* in our schema is thus the + or − residue in Z_1 which proceeds upwards as the message.

The schema outlined incorporates two physiological assumptions:

1. Different impulse sequences can mutually amplify or cancel one another (addition). There are various indications that this occurs. One example in the area of motor activity is the superimposition of locomotor rhythms of different frequency in relative coordination (von Holst, 1935f), where two rhythms can summate or cancel one another according to their phase relationship. An example from the field of coordination of sensory data is the exact linear superimposition of statically and optically produced central activity differences and the similarly purely additive positive or negative interaction of surgically-induced activity differences in the postural centre of fish (von Holst, 1950a).

2. The efference from the lower centre leaves behind it a certain alteration in conditions as a 'copy'. This assumption is plausible *a priori* for higher centres and can nowadays be regarded as true even for the lowest centres. Recent investigations of action currents in the spinal cord, using antidromic (backward flowing) impulses in motor fibres, have shown that it is quite probable that normal discharge of a motor ganglion cell, apart from passing through the efferent fibres, also spreads through the small dendrites ramifying in the vicinity and hence produces an alteration in conditions within the neighbouring intermediate neurons. Tönnies (1949) refers to this process as 'central feedback' and ascribes to it great importance in the control of excitation in the spinal cord. For our purposes, it is sufficient that one can conclude that our assumption is physiologically plausible.

---

[1] It should be emphasized that this 'negative feedback' is not a necessary component of the reafference principle and that it should not be confused with the latter! The decisive point in the principle is the mechanism distinguishing reafference and exafference. This distinction plays no part in cybernetic technology.

Application of the reafference principle to different neuromotor systems

EYE MOVEMENTS

We can now test the reafference principle to see which phenomena not explained by the reflex theory can be correctly predicted. In order to make use of perception as a source of information, let us begin with the human being, taking the optical system. We can assume that simple relationships exist in this instance, since the eye – which is a protected sphere located in the head – will not be adapted for mechanical disruption of its movement.

Reafference from the actively moved eyes could have two sources: (1) the retinal image displacement and (2) impulses from the receptors in the eye muscles. Only the first source is accessible to conscious perception; the participation of the second can only be deduced. Let us set out from an assumption which is critical, since it is improbable *a priori*. When the eye is immobilized and the muscle receptors are excluded (Fig. 3.5a), following the command 'visual sweep (eyeball rotation) to the right!' (taking all directional indications relative to the experimental subject), the efference copy must return upwards from the lowest centre in full force as a 'message', owing to the lack of any reafference from the retina or the muscles. Further, this 'message' must be the same as that which normally produces an environmental shift of the same size and in the same direction with the resting eye (Fig. 3.5b).[1] 'The entire environment has made a leap (of the same size as the intended eye movement) to the right.' This prediction is actually borne out! It has long been known from people with paralysed eye muscles that any intended movement produces perception of an evidently quantitative environmental shift in the same direction, and Kornmüller (1932) has provided more detailed verification of this by carrying out experiments on himself, following anaesthesia of his eye muscles. This 'apparent movement' cannot be distinguished from 'objective' perception of movement, as is to be expected, since according to the reafference

---

[1] This must be fulfilled, since with this environmental shift a retinal image displacement occurs opposite to that which would emerge with a completed visual sweep to the right. In our schema, the direction inversion signifies a change in sign of the afference from − to +, so that a + message equivalent to the + efference copy must pass upwards.

hypothesis the *same* message arrives in both cases. Thus, in this experiment there is, so to speak, *direct visual recognition of the efference copy.*

Following Hering (cf. ˙Trendelenburg, 1943, p. 240 et seq.), this phenomenon is interpreted as 'displacement of attention' in gaze movements. So far, a plausible physiological explanation has been

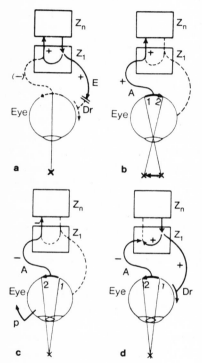

Figure 3.5. Explanation of perception of movement by the eye under normal and experimental conditions, using the reafference principle. The eye is in primary position; viewed from above. Z_1 = lower optical centre; Z_n = higher optical centre. The efference (E) passing to the eye musculature and the corresponding afference (A) produced by displacement of the image on the retina are indicated as in Fig. 3.4. In *a*, the immobilized eye receives the command impulse to glance to the right (turning impulse, Dr, for rotation to the right). In *b*, the viewed object (x) objectively migrates to the right, moving from 1 to 2 on the retina. In *c*, the eye is passively (mechanically) turned to the right, as indicated by the arrow p, and the immobile x moves from 1 to 2 on the retina. In *d*, the eye performs an active (commanded) gaze movement to the right (turning impulse = Dr, as in *a*), and the immobile x moves from 1 to 2 on the retina. (Further details in text.)

lacking. We shall see shortly that 'attention' has nothing to do with this, since the same phenomenon is found with non-conscious eye movements. Perception is, for us, simply a comfortable indicator for an otherwise relatively inaccessible physiological process.

The same perception of movement as that described above – though in the opposite direction, to the left – is obtained when the eyeball is passively rotated to the right with a pair of forceps (Fig. 3.5c). In this case, the movement command is lacking and the retinal exafference passes upwards unhindered as the message, once again producing 'pseudoperception': 'the landscape jumps to the left'.

If we now combine the first case with the second, that is, by passively turning a paralysed eye to the right in synchrony with the movement command (or by making a gaze displacement to the right with the *intact* eye, which is of course simpler), then two mutually complementary streams of impulses are actually produced (Fig. 3.5d): an efference copy which indicates movement of the environment to the *right*, and an exafference indicating movement to the *left*. Since these two streams cancel one another out at a low level (Z_1), there is *no* ascending message and neither stream of impulses is perceived. As we all know, the environment remains immobile, which in *this* case is also 'objectively' true. The 'correct' perception proves to be the sum of two opposing 'false' perceptual processes.

As with all technical constructions, this central apparatus has noticeable limits of accuracy. It only functions reliably in an intermediate visual range and with moderate speeds of movement. For example, if the gaze is turned markedly to the right and then rapidly passed up and down a vertical junction in a room, the latter undergoes pronounced 'apparent rotation' (Hoffmann, 1924). This is based (according to our interpretation) on the fact that the efference copy in this movement (following Listing's Rule) of a rotating eye only partially cancels out its afference, such that a message is passed upwards. The landscape also appears to leap markedly in the opposite direction when the gaze is rapidly moved back and forth from right to left. In this case, the efference copy is apparently produced too slowly, such that a restricted message is passed upwards in the first instant.

The fact that one can produce pronounced apparent movements of the environment, either through passive movements of an otherwise intact eye or through intended gaze movement with a mechanically fixed eyeball (a fact which was already known to Helmholtz), shows that the afference of the muscle receptors can only be of restricted significance (if at all) for our approach. If these receptors were always capable of indicating the position of the eyeball, as is the case with receptors in the limbs (see pp. 160–4), then apparent movements could only appear following their exclusion (Kornmüller's experiment). The role of these muscle receptors has apparently been so greatly overestimated – as we shall see in other instances as well – because the reflex theory does not provide for any concepts involving an intra-central feedback process.

The reafference principle is valid not only for so-called 'voluntary' movements, but also for those of an involuntary kind in which the eye 'scans' the visual field by sequentially fixating one point after another. This 'scanning' occurs both in active movement of the eyes (e.g. in reading a book, where every line requires four to five steps) and in turning of the head and body (labyrinth nystagmus) or presentation with a travelling landscape (optomotor nystagmus). In all of these cases, we do not usually perceive anything of the stepwise displacement of the image across the retina. Instead, we perceive the environment as travelling smoothly in one direction or (in the case of reading) as remaining stationary.

An interpretation of this phenomenon in terms of reflex physiology runs as follows. If the landscape passes by our eye (e.g. when looking out of the window of a train), the eye is initially carried along 'reflexly'. The consequent movement of the eye muscles indicates – through muscle receptors – the speed and direction of the environmental movement, since retinal displacement (which could also possibly act as an indicator) is lacking or negligible. When the muscular tension has reached a maximum, there is a further 'reflex' motor impulse in the opposite direction (the rapid nystagmus phase). The afference of the resulting, opposing displacement of the image across the retina is 'inhibited' or fails to reach the conscious level because of its rapidity.

Among other things, this interpretation contradicts the fact that one can produce an after-image on the retina by fixation of a bright cross and that this after-image, which persists in the dark, can be seen to wander alternately first in one direction (slowly) and then in the opposite direction (rapidly) with labyrinth nystagmus (which can be produced at will after practice). Thus, the image is *not* extinguished in the rapid phase (Fischer, 1926). As will be demonstrated shortly, this behaviour would be predicted by the reafference principle.

The much-discussed nystagmus apparatus can best be understood when the term 'optical scanning apparatus' is taken quite literally. Consider, for comparison, the locomotor apparatus of an arthropod (*Geophilus*; von Holst, 1933) following separation of the higher centres (Fig. 3.6). Through continuous (e.g. electrical) excitation of the ventral nerve cord, the locomotor system can be activated such that the legs perform coordinated walking movements, just as a uniform stream of impulses from higher centres sets nystagmus in operation. In addition – as with nystagmus – it can also be activated by unidirectional movement of the 'fixed landscape', by pulling the remaining fragment of the animal across the substrate at varying speeds. The legs *actively* accompany this latter movement: even those legs which are in the act of swinging forward will exactly follow every alteration in speed imposed upon the legs resting upon the substrate. The propulsive phase corresponds to the slow nystagmus phase: steered by the exafference, the legs and the eyes fixate the substrate. The swinging phase corresponds to the rapid nystagmus phase: the legs and eyes suspend the fixation and perform a pace in the opposite direction. In neither case do the higher centres need

Figure 3.6. A small fragment of an arthropod (*Geophilus*) which is drawn across the substrate with a hook attached at the anterior end. The legs raised from the ground are actively swung forward (→) in such a manner that each leg is placed exactly in the track of its predecessor and the resulting, overall pattern resembles the track of a bipedal animal.

to receive information from the individual paces. Just like the continuous stream of impulses from above, there is a reverse stream passing upwards, indicating the relative speed between the subject and the environment. The pacing apparatus fails when the moving substrate is unstructured (e.g. *Geophilus* on a mercury surface, or the eye passing over a homogeneous visual field), or when the relative movement becomes too rapid, in which case the legs and the eyes remain 'blocked' in an extreme backward position.

Let us now try to give a more exact interpretation of this scanning apparatus. The landscape begins to move past the eye and there is a resultant retinal displacement which immediately passes to the higher areas as a + message, bifurcating to Z_2 and the pacing centre Sz (Fig. 3.7a). This + message, in turn, at once descends from Z_2 as a + command and evokes a movement of the eye. The image is consequently returned to its previous position on the retina, producing a − afference which cancels out the + efference copy in

Z_1, so that the eye movement would be arrested (Fig. 3.7b) if the process in Fig. 3.7a did not immediately reappear. The movement thus oscillates in an equilibrium condition (Fig. 3.7c). The eye moves

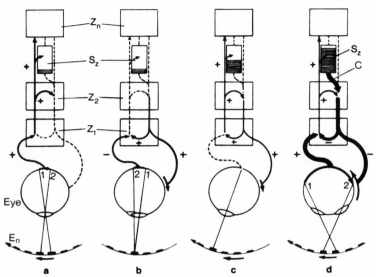

Figure 3.7. Explanation of the scanning movement of the eye in the example of optomotor 'nystagmic' movement accompanying movement of the environment (En). The schema follows those in Figs 3.4 and 3.5. Z_1 and Z_2 are lower motor centres; S_z is the scanning centre responsible for the rhythm; and Z_n is a higher optical centre. In *a*, the eye is at rest, the environment is slightly displaced, and an (arbitrarily selected) element of the visual image wanders across the retina from 1 to 2. In *b*, active accompanying movement has started, and the given element of the image has wandered back to its former position on the retina. In *c*, the eye moves with the same angular velocity as the environment and the image remains immobile on the retina. In *d*, the scanning centre S_z discharges and produces a powerful, opposing motor command in C. During the time elapsed with the rapid movement phase, the environment continues to move, and an arbitrary element of the image thus passes from 1 to 2 across the retina. The thickness of the arrows indicates the intensity of the streams of impulses. (Further explanation in text.)

so rapidly that there is no further retinal displacement (or almost none), and the entire feedback system maintains itself.

The scanning centre, S_z, is thus charged up until it abruptly discharges (acting as a 'ratchet system'[1]) and returns the eye to its starting position (Fig. 3.7d). This is the rapid phase, during which the

[1] A. Bethe (1940), in particular, has performed important experiments on physical ratchet oscillations as a model for physiological rhythms.

efference copy and the afference cancel one another out, apart from the small *ex*afferent residue which is brought about by the movement of the landscape and which continues on its upward path as a message independent of the phase-change. The same set of events is then repeated.

One can see that the entire apparatus is *not* a 'reflex mechanism' since, with sufficient auto-excitation or stimulation from above, it must pass over to autorhythmicity following *removal of all afference* (cf. Fig. 3.7c!). This is actually observed under experimental and pathological conditions.

ACCOMMODATION

We shall now turn from eye movements to another process: that of *accommodation*. At rest, the eye is set for long-distance vision, since the elastic lens is flattened by suspensory fibres. Proximal accommodation is ensured by a circular muscle which operates antagonistically to this tension and allows the lens to swell. As is well known, this apparatus – along with other criteria – operates in association with the retina to permit perception of the approximate distance and size of a viewed object. The retinal image, when uniformly focussed, is perceived as large and distant or small and nearby, according to the accommodation setting. Or, to use the usual formulation, the 'subject size scale' depends upon the 'imagined distance'.

The reflex theory can only explain this fact through the assumption that receptors in the accommodation apparatus, according to its setting, 'reflexly' exert a reducing or magnifying influence upon the image which is transmitted upwards. However, this possibility is discounted by the well-known fact that when the accommodation mechanism is paralysed with atropine, the – unsuccessful – attempt to fixate nearby objects makes everything appear to shrink in size (micropsia), although the incapacitated peripheral mechanism is naturally unable to elicit any 'reflexes'. Conversely, everything appears to become large (macropsia) when one attempts to fixate objects some distance away following tetanic contraction induced with eserine. The reafference principle explains this outcome.

Let us set out from a simple case (a), producing a sharp after-image of a nearby cross on the retina and then gazing at a wall some way off. The command for adaptation to distance produces in the lowest centre the corresponding efference to the musculature, together with its accompanying efference copy. Since the image of the cross on the retina remains sharp and unaltered, no reafference appears; the efference copy proceeds upward as the message and indicates a *positive perceptual process*, which (as is well known) reads as follows: 'The cross is (now) much larger.' We now proceed to (b), directly regarding first a large cross and then a small one at the same distance from the eye. In this case, an accommodation command is lacking; the exafference – as a result of absence of any efference copy – proceeds unimpeded upwards as the message and indicates a positive perceptual process: 'The (second) cross is smaller.' Following this (c), we take the larger cross, look at it close by and then move it just far enough away so that its image on the retina is exactly as large as the previous image of the small cross (in b). It now emerges that there is *firstly* the already described message from (a) stating 'the cross is larger' and *secondly* the message from (b) stating 'the cross is smaller'. Both the efference copy and the exafference cancel one another out in a lower centre, and *no* message passes upwards. Consequently, the perceptual mechanism must conclude: 'The cross has remained the same size'; which is exactly what it does do! Once again, the 'correct' perceptual process originates from two compensatory 'false' perceptual processes.

With this interpretation we can also easily understand the above-mentioned effects of micropsia and macropsia. When the peripheral musculature is unable to follow the efference, the retinal reafference must be absent, and the efference copy which passes upward as the message must indicate to the perceptual mechanism a size alteration in the predicted direction.

Once again, it is interesting that the reafference and the efference copy only cancel one another exactly within certain limits. Where accommodation no longer takes an active part – i.e. far away or very close by – one sees objects become smaller or larger, as is to be expected. The apparatus is also unable to follow rapid alterations in distance, such that an object which approaches rapidly appears to increase in size.

The principle set out here for the distance adaptation of the individual eye can be applied in just the same way to active binocular distance measuring devices – the convergence of the eyes, which increases with the closeness of the fixated object. This will not be

159

discussed in detail here since we cannot yet reliably state whether (and to what extent) reafference from the eye muscles is involved in the process.

LIMB MOVEMENTS

Let us now consider another apparatus: the moving limb. Here, more complex relationships are to be expected, since – in contrast to the protected eye – passive (mechanical) alteration of posture is possible in many respects, and the CNS must receive information about the latter in order to operate in an appropriate manner. Perception involves distinction of at least four different qualities of mechanical influence: touch, pressure, tension ('force'), posture. The first two are well known to everybody, and they overlap with one another. The fact that tension and limb posture are sharply distinguished can be clearly experienced when the intensity of the mechanical field is increased in a centrifugal field. If one's weight is more than doubled in this way, raising of the arm (for example) requires an astounding quantity of perceived force, although no external pressure is imposed upon it. Nevertheless, one is correctly informed about the momentary arm position or movement, and no surprise is encountered when the arm is examined following movement to a required posture.[1] This indication of posture is primarily given by receptors in the connective tissue external to the muscles; for measurement of tension there are further receptors both in the muscle fibres themselves and in the tendons. Together, they provide the so-called 'kinaesthetic sense'.

One can at once begin with a concrete question. How would a muscle respond to external loading if its centres incorporated interconnection of the efference and afference of the tension-measuring elements in the tendons, following the same pattern as that found in the postural mechanism of fish? A mildly stretched muscle is extended so that the tension increases (Fig. 3.8a). The increased afference (exafference) passes upward to Z_2 and indicates the size of the preceding efference; the muscle actively relaxes. When the muscle is unloaded, the converse will occur; the + efference copy passing from Z_1 to Z_2 increases the overall efference: the muscle activity contracts (Fig. 3.8b).

[1] Based on observations on myself, conducted in an encapsulated experimental chamber rotated by a kind of centrifuge.

160

The limb, however strong the muscle tonus may be, will 'spastically' follow the movement imposed from the outside. Mechanisms of this kind are widely distributed and they occasionally occur in quite pure form in pathological cases. An example in this direction is again provided by an operated *Geophilus* (see Fig. 3.6) which is

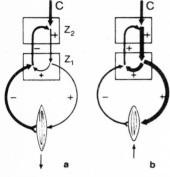

Figure 3.8. Explanation of the behaviour of a loaded muscle whose central apparatus is so constructed that – in analogy with Fig. 3.4 – it will actively adapt to supplementary passive stretching (*a*) and shortening (*b*), while the tonus determined by a higher command (C) remains the same.

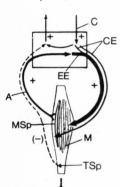

Figure 3.9. Behaviour of a muscle (M) mildly stretched through a higher command (C), whose central apparatus is so organized that it actively opposes any additional external loading (↓). CE = efference produced by the higher command, passing to the muscle and the muscle spindle fibre; EE = supplementary muscle efference which is directly produced by the afference of the stretched spindle; MSp = muscle spindle; TSp = tension receptor in the tendon ('tendon spindle') with a high stimulus threshold, which leads to the central apparatus illustrated in Fig. 3.8.

actively following an imposed speed of movement. Exactly the same thing can be seen in vertebrates (dog; toad) in which the spinal cord has been transected at the thoracic level; when the substrate is pulled away backwards, the back legs begin to move.[1]

With an actively innervated muscle of an intact warm-blooded animal, the response to loading is, of course, usually opposite in nature: there is contraction to an extent which will balance the load. This much-investigated *endogenous reflex* (P. Hoffmann *et al.*) passes

---

[1] Nevertheless, supplementary locomotor activity is necessary for the forward movement (swinging phase of the legs).

from the sensory neurone directly to the motor neurone and therefore operates without participation of an efference copy and without 'computation' in a higher centre. Its receptors are the muscle spindles, fine contractile fibres (representing about 1 per cent of the transverse area occupied by the remaining fibres with exclusively motor connections) which have both motor and sensory connections. If we imagine the afference from the muscle spindles as passing directly to the motor neurone leading to the remaining mass of the muscle, we can understand the operation of the endogenous reflex (Fig. 3.9).

If the *unloaded* entire muscle contracts or relaxes, the spindles remain silent, as long as they and the other muscle fibres alter their lengths to the same extent. If the resting or contracting muscle is loaded, the spindles are stretched as well, and they 'fire off' until they are discharged by contraction of the remaining muscle fibres.[1] Thus, the CNS determines the required position or movement, and the endogenous reflexes maintain this against external resistance. The tension receptors in the tendons, which have a higher stimulus threshold, may be ascribed with the function of switching off this reflex mechanism when there is inordinate loading and switching on the above-described adapting mechanism, so that the muscle does not tear apart.

With this refined trick, Nature has ensured that the organism maintains its equilibrium simply through muscular auto-regulation. If an impact from the left causes an animal standing at rest to sway to the right, the extensors of the right legs will stiffen as a result of increased loading, before the labyrinth has time to come into operation. An undisturbed animal standing or walking with mild resilience receives a solid support just at the moment of disturbance and only in the affected place. One can imagine the enormous saving in higher command impulses and muscular energy which this entails.

Thus, we can see that the reafference principle is apparently replaced by a differently operating *peripheral* mechanism in the endogenous reflexes, but that it is otherwise also operative in limb movements. Let us now make use of perception in order to follow this question somewhat further, as we did with the eye.

[1] We owe thanks to Dr Lissmann, Cambridge, for valuable data on proprioceptor function (verbal communication) taken from sources inaccessible to us, and to Prof. P. Hoffmann, Freiburg, for providing important literature.

At first sight, the well-known fact that disruption of the afferent pathways (dorsal roots) does not lead to perceptual illusions when the arm (for example) is moved, despite the lack of reafference, seems to disprove the reafference principle. However, the argument does not hold up. Any possible perception of apparent movement of any object is dependent upon a perceived object, so it will in this case be bound to touch and pressure afferences, which are similarly lacking. Similarly, in a quite analogous fashion, the feedback message from the efference copy of an eye actively moved *in the dark* of course does *not* lead to optical 'apparent movements of non-perceived objects'. A human being with an amputated arm does indeed continue for some time to dispose of the appropriate central representation; he can open and close the missing 'phantom hand', and he can say in which direction it is moving.[1] However, as many statements indicate, all this takes place in an 'imaginary sphere' which literally penetrates the 'real' (i.e. afference-induced) sphere.

Let us therefore test our question instead on the basis of concrete predictions which the reafference principle can make for perception.

If the 'kinaesthetic sense' of a limb is reduced, then (apart from the extinction of the endogenous reflexes and the associated emergence of muscular weakness) active movement on a solid substrate must produce a *positive* perceptual process. As a result of inadequate reafference from the tension receptor, the efference copy must proceed upwards as a message in association with any movement which loads the limb: 'The contacted object is moving away.' This prediction is accurate! For example, when the kinaesthetic sense of the extensor muscles of the legs is markedly reduced, as with polyneuritis, one has the impression (for instance, when stepping down from a stool onto the floor) that the substrate 'moves downwards elastically.[2] Presumably, the same interpretation can be applied to similar reports concerning *tabes dorsalis*, indicating that the floor is 'like rubber'.

As the motor counterpart to this effect, one should expect *over-extensive* limb displacement in active movement. As has already been explained for postural orientation, reduced reafference is accompanied by lack of what we have referred to as a 'peripheral stop',

[1] Unfortunately, there has apparently been no investigation to settle the important question as to whether motor impulses also pass into the stump of the arm.
[2] Personal observation reported by K. Lorenz (verbal communication) and others.

using technical terminology. In the lowest centre, the normal equilibrium state is not achieved, the efference continues and the limb movement becomes too extensive. The exaggerated, excessive movements of an attactic *tabes*-sufferer actually demonstrate this quite clearly.

To take another example: According to the reafference principle, it should make a great difference in perception whether pressure differences on the sole of the feet are produced by active movement or by a surface passively pressed against the sole. In the first case, one should notice no pressure differences, whilst such differences should be perceived in the second. D. Katz (1948) has found that when one stands up and actively loads the sole of the foot to differing extents by displacement of the body weight, by supporting the arms, by performing knee-bends, and so on, the perceived *difference threshold is twenty times higher* than when one lies on one's back and a corresponding pressure is exerted upon the sole.[1]

Thus, perception once again demonstrates the validity of the reafference principle. The mechanism in this latter case is doubtless to be found in a fairly high central area, since the behaviour of the four limbs is taken into account in one framework.

REAFFERENCE AND LOCOMOTION

We can consequently measure the extent of reafference incorporated *as an integrating component* in a neuromotor apparatus according to the degree of exaggeration and inexactitude of the movement following destruction of the sensory pathways. In this respect, some interesting differences are observed: in lower swimming and crawling forms (fish; amphibians), locomotion is still completely normal following deafferentiation; in walking mammals, it is extremely atactic, and the complex motor sequences of the human hand can only be performed as individual, greatly exaggerated and disorganized fragments (Foerster, 1936). From this, it follows that with the simple, monotonous movement patterns the CNS, in principle, performs everything independently and thus has an 'automatic' character (von Holst; P. Weiss), whilst higher motor patterns *do not need* '*reflex stimuli*', but *do apparently need reafference*! The series from

[1] The author himself attempts to explain this fact in terms of a 'Gestalt interpretation of the entire range of bodily experience'.

swimming through crawling, running, climbing and grasping to touching (hand; tongue) passes from movements which are adapted for no afference, through those which require reafference and ex-afference to movements which are primarily adapted for exafference.[1]

The old controversial question: 'Is movement of the limbs a reflex or an automatic process?' can, in this light, be set aside. The alternative was false! To give a pictorial illustration: the locomotion of a swimming fish takes place blindly into the dark, whilst the moving hand needs an illuminated environment. A deafferentiated hand is like a blind man, who cannot go on his way since he does not know where he is. This does *not* mean that the central motor drive for the hand is weaker than that of a swimming fish. And exactly as the sense of touch guides the blind man, the eye aids a sensorially paralysed hand: both move around much better with such auxiliary afferences.

THE INTERACTION OF SEVERAL AFFERENCES

The situation in the CNS becomes somewhat more complex where afferences from different parts of the body which can be moved in opposition to one another are involved in the posture and movement of the entire animal. A simple example: In arthropods, the direction of running is determined by ganglia in the head. If the right oeso-phageal commissure is transected, thus eliminating operation of the right sensory centre (supraoesophageal ganglion), there is a resultant turning tendency – analogous to that found with postural orientation following unilateral suppression of the vestibular nucleus – which in this case produces movement to the left. The animal flexes towards the left when moving forwards and to the right when moving back-wards, thus producing circles to the left in both cases (von Holst, 1934).

If we make the justifiable assumption that the command displac-ing the tonus in the segments originates from a subordinate centre (suboesophageal ganglion), it is to be expected in accordance with the reafference principle that reduction in the number of segments will increase the curvature exhibited by the remaining segments. This is because the loss of reafference, which produces a further

---

[1] Like the expressive musculature of the face, the tongue does not possess proprioceptors; in the fingers, the proprioceptors are the most important providers of exafference.

imbalance in level in the higher command centre, can only be countered by a corresponding increase in the range of movement, producing additional reafference. This expectation is borne out in practice:

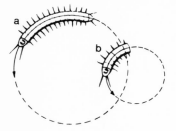

Figure 3.10. Arthropod (*Lithobius*) following transection of the oesophageal commissure on the right side. It describes circles with a specific mean diameter. *a*: otherwise intact animal; *b*: animal shortened posteriorly.

if a many-limbed arthropod (e.g. *Lithobius*) is shortened by removal of the posterior segments, the curvature increases with increasing reduction in length (Figs 3.10 and 3.11), such that half of the animal will describe circles which are scarcely half as big as those produced by the entire animal (von Holst, 1934). This is a curious fact in the light of the reflex theory!

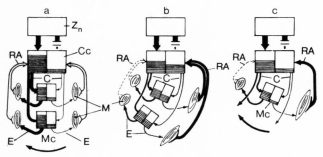

Figure 3.11. Schema to explain Fig. 3.10. Zn = higher sensory centre (supraoesophageal ganglion); Cc = command centre (suboesophageal ganglion); C = command to the lower motor segment centres (Mc), of which only two are represented; E = efference to the muscle, M; A = afference; R = reafference from postural indicators in the joints of the body. The thickness of the arrows indicates the intensities of the streams of impulses. The right stream from Zn to Cc has been disrupted. In *a*, the asymmetrically charged command centre produces an equivalent asymmetry in the motor centres, and the efference of the latter produces body curvature (tonus displacement) to the left (lower arrow). In *b*, the reafference of this curvature charges the command centre in a balanced manner and produces an equilibrium between the required asymmetry of innervation and the asymmetry in reafference. In *c*, after removal of the second motor centre the reafference is once again inadequate, and there is further disequilibrium which necessarily leads to an increase in tonus displacement (lower arrow).

It is also to be expected that in an intact animal which is moving in a straight line, *passive* deflection of the posterior end (for example) will lead to an *active* counter-deflection of the anterior end – as long as other disruptive effects are absent – since the exafference from the posterior end must be compensated by an opposing afference from another area in order to produce, *in toto*, the correct overall afference on the way up to the higher centre. This behaviour is also well known for many arthropods under the name of the 'homostrophic reflex' (Fig. 3.12).

We can now turn to a more complex case, that of postural orientation in higher vertebrates and man. In the fish, postural orientation can be easily surveyed, since the static sensory organ is rigidly incor-

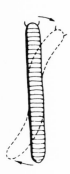

Figure 3.12. Arthropod (*Iulus*) in which the posterior end is passively deflected to the left and the anterior end is subsequently actively turned to the right ('homostrophic reflex').

porated in the body. In addition, the motor system has no additional task to perform, whatever the required posture, since the mechanical equilibrium is of no importance. By contrast, *our* upright posture is labile and every requisite posture necessitates special innervation relationships. Furthermore, the head, body and limbs can be moved relative to one another. In active or passive inclination of the head, the labyrinth *must not* elicit any 'limb reflexes', since the latter would only jeopardize bodily equilibrium! Let us look at the behaviour which actually occurs (Fig. 3.13a–d).

The head and body of a resting animal are passively inclined to the right: a realigning compensatory movement is made with the limbs (Fig. 3.13b). If the head is immobilized in its spatial location and only the body is inclined, the same movement appears (Fig. 3.13c). If the head alone is inclined, the attempt is made to restore it to an upright position, but the body remains immobile (Fig. 3.13d).

These observations can most easily be understood if it is assumed that the postural receptors are located in the body, as has often been suggested (Trendelenburg, 1906, 1907; Fischer, 1926). However, the behaviour of the eyes does not fit in with this at all. In (b), the eyes are rotated slightly to the left (relative to the head), in (c) they are rotated slightly to the right, and in (d) pronounced rotation to the left occurs, so that they once again exhibit roughly the same position to the body. This is entirely analogous to the orientation of the head and the posterior extremity with the 'homostrophic reflex'.

The simplest explanation of the overall behaviour runs as follows:

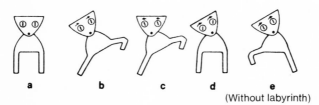

a   b   c   d   e

(Without labyrinth)

Figure 3.13. Sketch explaining the correcting movements of the limbs and eyes of a warm-blooded animal (mammals; birds) accompanying passive changes in posture (direction of inclination indicated with respect to the animal). *a:* normal posture; *b:* head and body inclined to the right; *c:* only the body inclined to the right; *d:* only the head inclined to the right; in *e*, there is the same situation as in *d*, but after removal of the labyrinths (cf. Fig. 3.14c).

two afferences (at least) are involved in the posturing of the body, head and eyes; one originates from the statoliths in the head and the other comes from postural receptors in the neck muscles. These two streams of impulses are mutually subtracted in respect to the posture of the body and added in respect to the direction of the eyes (Fig. 3.14a). If this interpretation is correct, one should – according to the reafference principle – expect a quite specific functional disruption as soon as the afference from the two sides of the neck is artificially brought into imbalance. The CNS would evaluate this intervention as 'inclined body relative to an erect head' and produce an appropriate compensatory movement of the limbs. This actually proves to be the case. If a cold, wet compress is applied to the left side of the neck below the mastoid area and a hot compress is applied symmetrically on the right (thus producing a slower impulse frequency from the receptors on the left side and a faster one on the right), the expected postural alterations of the limbs occur.[1]

The reafference principle also requires exactly the same misinterpretation when the head is actively or passively inclined following

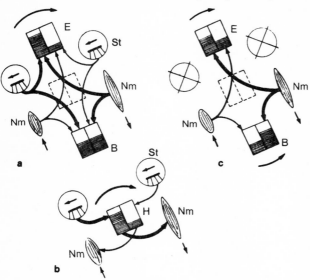

Figure 3.14. Schemata (analogous to Fig. 3.3) explaining the behaviour of a warm-blooded animal with an intact statolith apparatus (*a, b*) and following removal of the latter (*c*), when the head is (passively) inclined to the side (cf. Fig. 3.13d, e). *a* and *b* equally apply to active inclination of the head. Postural centres: B = for the body; E = for the eyes; H = for the head. Nm = neck muscles. In *a* and *c*, the head postural centre has been omitted in the interests of clarity; it is represented in *b*. The extent of afference from the statolith apparatus and from the postural receptors of the neck muscles is indicated by the thickness of the arrows. Unequal charging of the eye postural centre (E in *a* and *c*), the head centre (H in *b*) and the body centre (B in *c*) in each case provokes a turning tendency indicated by the curved arrow. (For further details, see text.)

suppression of both labyrinths. In this case, too, since there is no further information regarding inclination of the head, the CNS must 'believe' that the head is erect and the body inclined, consequently proceeding to correct the apparent inclination of the latter. This has

[1] The phenomenon was first described by Griesmann (1922) and confirmed by Fischer and Wodak (1922). Goldstein and Riese (1925) used it as an argument for a general theory of plasticity. As far as we are aware, no physiological interpretation has been presented.

also been known for some time as an actual occurrence (Dusser de Barenne; Magnus, 1924; cf. Figs 3.13e and 3.14c).

Although 'neck reflexes' of this kind have been known for some time – they were discovered by Barany – it has usually been assumed, from the fact that we can incline the head in all directions *without* 'reflexes' being exhibited by the body, that such neck reflexes are typically lacking in man. This widespread accepted opinion is once again a probable outcome of the reflex theory, which dictates that a 'stimulus' must always be followed by a movement – a movement which is in this case lacking.

The described extinction of the labyrinth and neck afferences only applies to the stream of impulses descending to the limbs on the trunk; it does not affect the eye where the two components are additive. Inclination of the head to the left produces rotation of the eye to the right (as can be easily observed with a mirror), which is exactly the same as that produced by inclination of the entire body without movement of the neck and by bending of the neck whilst the head is maintained erect (i.e. turning of the body to the right).[1] This effect was already known to Magnus and his co-workers. This summation ensures that, when the head is moved, the eyes retain their position relative to the vertical plane and thus maintain their visual field to some extent.

As far as the head itself is concerned, the observed behaviour is easily understood. The centre for head posture receives its afference exclusively from the labyrinth and transmits its efference to the neck muscles. The head together with its neck muscles is comparable to the entire musculature of the fish (Fig. 3.14b). When passively inclined, it returns to the erect position, and it can also be actively inclined (command from higher centres). In both cases, the reafference flowing to the centres for body and eye position from the labyrinth and the postural receptors in the neck muscles is the same. Thus, we can understand that active or passive movement of the head makes no difference for the body and the eyes. One can give equally plausible explanations for the fact that the body and the limbs can similarly be moved to various requisite positions, whilst the head independently maintains its spatial location.

Through this interaction between the neuromotor mechanisms

---

[1] The relative contributions of the two afferences, in this case represented as equivalent (Fig. 3.13), can also be unequal for control of the eyes.

one can understand why the organism behaves *as if* the 'postural sense' were normally located in the body and as though it were able to wander up into the neck following removal of the labyrinths, such that inclination of the neck would abruptly produce 'postural reflexes'. This really does pose a problem for the reflex theory!

Conclusions

This account represents a preliminary, rough outline. Nevertheless, it may clearly show why we believe that the reafference principle possesses an advantage over other interpretations. The reflex theory describes everything which is evoked by stimuli with the same term. This, in itself, is inoffensive – after all, we need collective concepts – but this term is underlain by the seductive reflex-arc concept, which almost always presents us with false explanations. The counterpart of the reflex theory – the *plasticity* theory (Bethe) – according to which everything is connected up with everything else and excitation can spread in all directions in the CNS as in a nerve reticulum, is doubtless justified in its negation of the reflex interpretation; but neither these two theories nor any other given theory of the CNS can predict in a concrete case what will actually happen – *and what happens is never arbitrary*.[1] The reafference principle provides concrete predictions which can be tested to show the validity of the principle and its limits. It represents *one* mechanism *among others* and does not prejudice the possible occurrence of automaticity, coordination and spontaneity. It would therefore seem appropriate to link the opposing interpretations one to another. Since the reafference principle explains in a uniform manner particular phenomena throughout the CNS from the lowest processes (passive and active positioning of the limbs, relationship between the different parts of the body) to those at a very high level (spatial orientation, perceptual processes, sensory illusions), it provides a bridge between the lower levels of neurophysiology and higher behavioural theory.

It has often been asked whether insects can distinguish their own

---

[1] Bethe himself is presumably in agreement with this interpretation. This is demonstrated by his various attempts to give a more detailed explanation of actual events, using physical models. In our opinion, his derived principle of mechanical 'transitional coupling' (Bethe and Fischer, 1931) represents a good model; the discussion on pp. 152–71 of the present paper is simple a detailed representation of transitional coupling.

movements from movement in their surroundings. Mathilde Hertz (1934) instinctively answered this question positively, but was unable to give a plausible explanation of the manner in which it occurs. Other investigators have followed von Buddenbrock (1937) in giving a negative answer on the grounds that the relative movement between the animal and the external environment is, of course, the same in both cases. We can now recognize the following fact: the *eye* is indeed unable to distinguish between its own movement and that of the environment, but the *animal* – which, after all, does possess a CNS consisting of more than connecting leads between the receptors and the muscles – is *well able* to distinguish the two. With the aid of the reafference principle, it achieves recognition of the *constancy of its own objective environment*.

Such *constancy phenomena* in fact play a major part in human psychology. Well-known examples are the perception of an immobile environment when the eyes are moving (spatial constancy) and the perception of an object as retaining a particular size independent of distance (size constancy). As we have seen, the reafference principle explains both of these phenomena. The fact that it does not explain other constancy phenomena into the bargain (e.g. 'colour constancy' of optically perceived objects) is not a drawback but an advantage of the principle, which – because of its concrete formulations – does not permit pseudo-explanations of heterogeneous factual matter.

Finally, the reafference principle also provides a quite specific contribution to the question of the *objectivity of perception*. We have repeatedly seen that the 'correct' information is simply the resultant of two 'false' bits of information, each of which possesses in its own right the character of 'correctness'. For a lower centre which receives only *one* afference, all information is 'correct' in the same way. The question whether a perceptual process can also be *'objectively'* correct, or whether it is only 'apparent', can only arise where several different afferences are combined. 'Objectively correct' then means no more than coincidence of different bits of information, and information is evaluated as 'apparent' when it does not fit in with other information. In this respect, the lowest centre is unconditionally stupid – but we must also consider the fact that even the highest centre can never be more intelligent than its afferences permit, and that every individual afference is 'fallible'.

It is hoped that this article will contribute to the gradual disappearance of attempts to describe the functions of the highest developed organ of the body with a few primitive expressions. The sooner we recognize the fact that the *complex higher functional Gestalts* which leave the reflex physiologist dumbfounded in fact send roots *down to the simplest basal functions of the CNS*, the sooner we shall see that the previously terminologically insurmountable barrier between the lower levels of neurophysiology and higher behavioural theory simply dissolves away.

4 The central nervous system *1956*

The comparative physiology of the central nervous system (CNS) has recently adopted an entirely different form in several respects. This is due partly to the progressive refinement of technical methods and partly to the combined emergence of new concepts and the slow extinction of traditionally accepted ideas. The following significant advances can be listed:

1. The ever-increasing refinement of techniques for recording action currents has recently permitted tracing of the activity of individual neurones right into the ganglionic mass and even allows penetration into the interior of the cell body.

2. Refined methods of stimulation within the CNS have, in a number of cases, led to the discovery of structures (pathways and 'nuclei') whose stimulation activates *natural* behaviour patterns. On the other hand, systematic analysis has in many instances dissected the complex behaviour of intact organisms into lower functional elements, such that the former fields of 'neurophysiology' and 'animal psychology' nowadays confront one another at the level of *behavioural physiology.*

3. The controversial discussion of 'the theory of centres' has gradually been overtaken by a more natural interpretation which involves numerous *functional systems* within the CNS, whose functional components are located in different layers of the CNS, and in each case link up with specific effectors and receptors to form *units,* the functional structure of which must be established in each particular case.

4. Even the old-established 'reflex theory' is beginning to disintegrate; the 'reflex arc' has repeatedly proved to be a fragment of such functional units. Only the term 'reflex' continues to survive in an indeterminate form and employed in various ways.

5. The theory of *automaticity* of central nervous elements – i.e. the concept that many neuronal systems continue to operate (on the basis of metabolic processes) even when the peripheral afferences have been suppressed – has been increasingly supported. It is supplemented by the fact that the majority of peripheral *receptors* also exhibit *automaticity*; that is, that even without an *adequate stimulus* a continuous *afferent stream of impulses* passes into the CNS. Specific stimulation of the receptor is merely indicated by a *change* in this afference. From this it follows (among other things) that transection of an afferent pathway signifies *more* than simple exclusion of peripheral stimulation.

6. Finally, there have been various successful attempts to penetrate into the physiology of the CNS (e.g. in psychology, psychiatry, sociology and so on) using mathematical and technical concepts labelled with the term 'cybernetics'. This has a pronounced 'lubricating' effect at least in those cases where conceptual patterns have been frozen into schematism.

It is not the task of this report, and those which are to follow, to provide information about the factual material accumulated year after year. Instead, even more than in previous years, the intention is to discuss individual themes whose treatment provides an adequately coherent picture – without illegitimately relying upon the fantasy of the reviewer. The present report is concerned with certain extremely important regulatory mechanisms associated with posture and movement, which – in the sense indicated above – represent typical conceptually (and functionally) separable functional systems. Their most conspicuous histological component is the complex receptor possessing both motor and sensory innervation, and we shall therefore begin with a discussion of this component.

The muscle spindle system of mammals (cat; rabbit)

Among the numerous early histologists who described the muscle spindles, particular prominence must be accorded to Ruffini (1893), whose description has been confirmed and expanded by Barker

(1948). Muscle spindles are particularly numerous in the extensor muscles of the limbs and somewhat less common in the retractors. In addition, they are found in particularly high densities in the muscles of the human hand and in those of the eyeball; they also occur in the tongue, in the larynx, in the diaphragm and even in

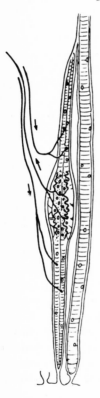

Figure 4.1. Structure of a muscle spindle (with only a simple sensory supply), following Barker (1948).

the *tensor tympani* (discovered in the sheep). A simplified diagram is given in Fig. 4.1. The important feature is that the spindle is not strung between the other muscle fibres, but lies parallel to them. The muscle fibres of the 'spindles' themselves posses motor innervation at both ends involving thin nerve fibres (ca. 3–7μ). In the middle, enclosed in a capsule containing lymphatic spaces, there is a richly nucleated, non-striped, non-contractile section which is surrounded by outlyers from at least one (usually several) sensory nerve fibres of large cross-section (15–17μ). In mammals, the motor nerve fibres of the spindle muscles hardly ever possess branches to other muscle fibres, but they do branch to other spindles. The remaining mass of the muscle is specially supplied by thicker nerve fibres (10–12μ).

Since, as is well known, the speed of conduction increases with fibre thickness, whilst the nerve cables are generally kept as thin as possible for reasons of economy, the thick afferent nerve fibre (speed of conduction 60–125 metres per second (m/s)) itself betrays the fact that it participates in a rapidly operating system. A similar indication is provided by the frequently confirmed finding that the

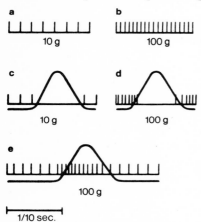

Figure 4.2. a and *b*: action potentials of an individual muscle spindle (cat) with different degrees of loading of the muscle; *c* and *d*: the same during contraction elicited by stimulation of the muscle nerves (spindle muscles not stimulated); *e*: a response corresponding to that in *d*, but taken from a tendon receptor. (Following Hunt and Kuffler, 1951; somewhat schematized.)

associated dorsal ganglion cells are *directly* connected with the motor cells of the same muscle (i.e. through only one synapse), providing the only known case of a 'monosynaptic reflex'. The thin efferent fibres of the spindle organs, which represent 30 per cent of all efferent fibres in extensor muscles, conduct two to three times more slowly (15 m/s or thereabouts) than the remaining efferent fibres (approx. 50 m/s), which themselves never possess branches to the spindle muscles.

More detailed accounts of the physiology of the spindles, which support and expand upon the main points of earlier interpretations (particularly those of P. Hoffmann), have been provided by Matthews (1933) and a number of recent authors (Kuffler *et al.*, 1951; Eccles *et al.*, 1954; Lloyd, 1953; Katz, 1950; Hunt and Kuffler, 1951a, b,

1952a, b; Granit, 1950–52; and others). The findings are predominantly based upon combined electrical recordings from individual fibres.

Under normal conditions, the *intact muscle spindle* with retained afferent and efferent supply is usually *persistently* active, even when there is no external loading of the muscle. The frequency of discharge increases with the loading of the muscle (Fig. 4.2a, b). Thus, in this respect, the spindle behaves just like the *stretch-receptors* in the *tendons*, which are supplied with afferent fibres of roughly the same thickness, but have no inherent motor innervation and generally possess a much higher stimulus threshold (Hunt and Kuffler, 1951b). The discharges from the spindle decrease or disappear when the *remaining* mass of the muscle is induced to contract by stimulation of the thick efferent nerve fibres (Fig. 4.2c, d). In this respect, the tendon receptor behaves in the opposite manner (Fig. 4.2e). Finally, just as with external loading, the spindle discharge frequency also increases when the thin efferent fibres leading to the spindle muscle fibres are stimulated (Hunt and Kuffler, 1951a, b; 1952a, b). Thus, the spindle always discharges when its middle, non-contractile section is deformed by pulling, whatever the cause of the pull. As with many other receptors, the discharges are in each case preceded by local depolarization, which sparks off the conducted impulses (Katz, 1950).

The discharges of the spindle simultaneously indicate two facts: (1) the momentary tension of the spindle muscle fibres, which corresponds to a specific discharge frequency in the spindle; and (2) the *change* in tension, which magnifies the frequency curve according to its speed. During the increase in tension there are 'too many' discharges, and during the decrease there are 'too few' (the two effects representing mirror images). Thus, just like so many other receptors, the muscle spindle is simultaneously a measuring instrument for 'proportional quotients' and 'differential quotients'.

As far as has been determined from measurements of activity in peripheral fibres and in the spinal cord (Eccles *et al.*, 1954), the following account can be given about the nervous pathways: The most important connection is the monosynaptic link leading back to the same muscle. Contraction of the muscle necessarily has an unloading effect on the spindles within it, and the spindles consequently discharge more slowly or cease to fire at all. In addition, the afference from the spindles passes through several synapses to further motor neurones linked to synergists in the same muscle, and these synergists are similarly activated, though to a lesser extent. Finally, the afference passes along pathways of some length to the antagonist(s), whose activity is reduced (inhibited). Stimulation of a *tendon* receptor, on the other hand, conversely produces inhibition of the same muscle and its synergists, whilst facilitating the antagonists.[1]

Apparently, activity of the spindles is not represented in the cortex. Stretching of a muscle (with the exclusion of tactile and other auxiliary stimuli) cannot be recorded in the cortex of the cat (Mountcastle *et al.*, 1952). Further, the spindles do not assist the sense of tension or 'force', nor do they aid determination of limb position. Where no inherent postural receptors are present, as in the case of the eye, passive mechanical displacement is not perceived as such. (This will be considered in a later report.) With the tongue, elimination of the skin receptors suffices to prevent perception of the position of the tongue, although spindles are particularly numerous in the anterior, mobile section of the tongue (Carleton, 1938; Cooper, 1953).

On the other hand, strong *influences* are exerted upon the spindle system of the limbs *by higher central areas*. The well-known extended posture of a decerebrated animal is the result of tonus facilitation of the spindle system (Sprague *et al.*, 1948). Electrical stimulation of the brain-stem (*formatio reticularis*) produces gradually increasing activation of the spindles, which persists after the stimulus (approx. $\frac{1}{2}$ min.). Less regular effects – i.e. effects which are dependent upon other factors as well – can be evoked from the mid-brain and the vestibular nucleus (Sprague *et al.*, 1948; Granit and Kaada, 1952). Direct stimulation of fibres in the pyramidal tract produces abrupt activation of the spindles which ceases at once when the stimulus ends (Granit *et al.*, 1952). Elicitation of certain reflexes, such as that of ear-scratching, also produces augmentation of spindle activity in certain limb extensors.

Thus, in all such cases the spindle muscles are innervated, the

---

[1] It should not be forgotten that some authors also report that with intense stimulation of the spindles there is an inhibitory effect which operates on the same muscle (Hunt and Kuffler, 1952); but this observation is disputed (Brock *et al.*, 1951; Hodes, 1953), and the reviewer believes it to be dubious.

spindles are consequently stimulated by stretching, and the entire muscle is therefore 'reflexly' induced to increase its activity. The diagram in Fig. 4.3 is intended as an illustration of the main connections.

From what has been said, it is evident that we are confronted (primarily) with a simple feedback system. The spindle afference is rigidly coupled with the muscle efference in a positive sense, whilst

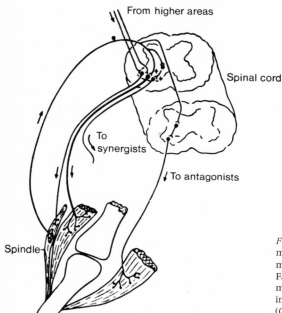

Figure 4.3. Schema of the main pathways in the muscle spindle system. Facilitating effects are marked with +, inhibitory effects with − (Constructed from Eccles *et al.*, 1954.)

the latter is in turn coupled with the stimulated condition of the spindles in a negative fashion. In other words: spindle activity evokes muscular activity, whilst the latter silences the spindles through unloading.

The spindles themselves can be stimulated in two (inherently distinguishable) ways: (1) by external loading of the entire muscle, and (2) through efferent impulses (e.g. a voluntary command from higher areas passing to the spindle muscle fibres). In both cases, the spindle induces its dependent muscle elements to contract until it is itself *un*loaded, i.e. until a quite specific length of the muscle is produced. However, this length can vary according to the efferent stream of impulses received by the spindle fibres at any given time. This leads to two important consequences:

1. No higher central area needs to be concerned with disruptive resistances and variations in loading which occur in the limbs, for example, in standing or running. *The muscle spindle system ensures that these effects are eliminated and that the momentary requisite posture or movement is 'successful'.* The tendon receptors probably come into operation only when inordinate resistance or inhibition is encountered, in which case they suppress further movement.

2. Any *voluntary command* – or, for example, an influence emanating from the labyrinth – fundamentally *only needs to pass to the spindle musculature; the remaining mass of muscle automatically follows through the action of the spindle system.* The fact that the entire musculature is nevertheless innervated by direct means as well may be primarily a phylogenetic consequence. In addition, this may be advantageous for particularly rapid responses; but the delay imposed by the feedback circuit of the spindle system is actually less than 1/100 s. In any case, the incorporation of innumerable muscle spindle systems in the overall musculature of the locomotor apparatus represents a *saving for the CNS*, the significance of which can scarcely be overestimated.

The spindle system in cold-blooded animals (frog)

All of this goes to show that the histologically simplest monosynaptic 'reflex arc', which is widely regarded as a typical, general and elementary functional unit, is in reality a *fragment* of a far from elementary system which is adapted to perform quite *specific* functions. Without a doubt, this system is not a primitive feature but something which evolved late in phylogeny. Investigations on cold-blooded animals (frogs) show that they possess a homologous system which is obviously at a lower level of development (Katz, 1949; Kuffler, 1953).

The frog possesses two separate neuromuscular systems, a fact which was histologically identified some time ago (Sommerkamp; Wachholder; P. Krüger) and which has since been verified physiologically. Apart from a rapid system, which approximately corresponds to that in warm-blooded animals, there is a slow 'tonic' system, whose muscle fibres only contract (locally and persistently) in response to multiple stimulation, and whose thin nerve fibres conduct about five times more slowly (2–8 m/s) than the thick fibres of the rapid system. The spindles are in this case equipped with muscle fibres which are similar to those in the slow system and which – an important feature – receive their motor innervation from nerve fibres which *also* innervate *other* muscle fibres. This means that the spindle muscles and the remaining mass of muscle cannot receive separate impulses, so that (for example) the spindles will also continue to fire when the loading muscle contracts.

Thus, in this case there can only be, at the most, a feedback system *in statu nascendi*. It is reasonable to suggest that the more highly developed system of warm-blooded animals has been derived from such intermediate stages.

It should also be mentioned that – as a purely convergent phenomenon – an organ corresponding to the muscle spindles also occurs in arthropods (in the caudal muscle of crabs). This organ, which was recently identified histologically (Alexandrowitsch), possesses both motor and sensory innervation, like the spindle, and the sensory terminal bundle surrounds a non-contractile section of the muscle fibre. In this case, too, the adequate stimulus is stretching (Kuffler, 1954). We do not yet know the details of the interconnections.

The postural receptor system

As one can see, the functions of the two types of receptor which respond to tension – the muscle spindles and the tendon receptors – are sufficiently clearly distinguishable in vertebrates. The tendon receptors have a protective function in overloading, and they are probably the exclusive agents of the 'sense of force'. Neither of these two types is involved in the recording of limb position (i.e. the momentary *alignment of the joints*). This is evident enough from the observation that in a constant centrifugal field ('roundabout'), where the raising of a limb is associated with a considerable increase in objective and subjective exertion, the position of the limb is nevertheless correctly perceived when the eyes are closed, although there is an accompanying lack of touch and pressure stimuli, and the tension is everywhere increased, thus predisposing to misinterpretation (von Holst). Consequently, limb position must be recorded by load- or tension-*independent* receptors of the kind found in the close proximity of the joint surfaces. This is supported by the observation that it is impossible for a human being to specify a passively determined finger position after exclusive suppression of the nerves in the joint (Stopford, 1921), and that the 'postural reflexes' of the neck are elicited by receptors in the intervertebral joints of the neck, rather than by muscle receptors (McCouch *et al.*, 1951).

A recent systematic investigation (Boyd and Roberts, 1953) has provided important information regarding the function of postural receptors in the knee-joint of the cat. Within the joint capsule, there are two types of receptors, each functioning in a distinct manner. One type (probably the Vater-Pacinian corpuscle) is silent when at rest and fires only when the position is changed. The other type (probably the Ruffinian corpuscle) exhibits *continuous activity* and responds to any position of the joint with a specific frequency of discharge. In addition, it responds to change in position with overshooting of the frequency curve, and it thus represents a 'proportional and differential quotient' measuring organ. The adequate stimulus is a specific deformation of the receptors. Sensitivity is high; for example, an angular change of 2° increases the frequency of a receptor from 13 to 16 discharges per second, and movement back to the original position decreases the frequency to 13. The different receptors respond to different directions of movement, in each case producing their greatest increase or decrease in frequency, such that every position corresponds to a specific pattern in the continuous afferent stream of impulses to the CNS.

Thus, it can be seen that the vitally important information concerning the position of the limbs is provided through a system of *excitatory patterns*, which are *continuously* present and whose 'appearance' changes in a regular fashion with any change in limb position. The two types of receptor in this case operate according to the same principle as those in the sensory terminals of the labyrinth (Lowenstein, 1950). It is thus comprehensible that the CNS is able to compute afferences from the labyrinth, the neck and the limbs such that the physically correct upright position (or a specific position departing from this) is tenable (von Holst and Mittelstaedt, 1950) and that

183

one can correctly judge the position of the parts of the body relative to one another and to the gravitational field.

In summary, all of these results show in the clearest possible fashion that electrical stimulation of a dorsal spinal cord root or a *bundle* of afferent fibres – which is so revered by 'classical' reflex physiologists – is an extremely coarse and unnatural intervention. Just as with direct stimulation of the optic nerve one obtains, instead of an organized excitatory pattern, a random turmoil! Only where the afferent cables are derived from the *simplest* receptors, or where intracentral pathways are conducting already computed, integrated simple data, does direct stimulation have *direct* results: well-organized orientation movements or instinctive motor patterns emerge.

5 The participation of convergence and accommodation in perceived size-constancy *1955*

It is well known that an approaching object which is viewed with both eyes is seen – within certain limits – to remain *approximately* constant in size. This, despite the fact that the size of the image on the retina increases in size in accordance with curve *a* in Fig. 5.1,

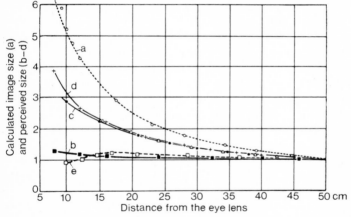

Figure 5.1. Graphical representation of the perceived size of an object in relation to its distance from the eye, with the size at a distance of 50 cm taken as unity (i.e. corresponding to a visual angle of about 8°).

a: size of the retinal image, as calculated (- - - -) and as measured from calibration of the after-image (o).

b, c, d: perceived object-size. *b*. with binocular fixation (■); *c*, with accommodation maintained constant (●); *d*, with convergence maintained constant at 40° (+) or 20° (△).

e: sum of the distances of curves *c* and *d* from *a*.

as has been demonstrated by reconstruction of the optical pathways and by direct linear measurements of after-images projected on to a calibrated scale. This phenomenon is referred to as the 'size-constancy of viewed objects' (Hering). The precision of this constancy function can be tested by altering the objective size of an approaching object such that its *perceived* size remains exactly constant. On the basis of the inverse values for the required size modifications of the object, curve *b* in Fig. 5.1 can be calculated for the constancy apparatus. As can be seen, this is close to the ideal of a horizontal straight line.[1] The difference between curves a and b is an indication of the enormous achievement of the correcting mechanism, which may itself depend upon (1) the mechanism for convergence and (2) the mechanism for accommodation. These two mechanisms are centrally coupled, but in some people they can be voluntarily separated so that the part played by each one can be investigated separately.

Convergence. The visual axes are crossed, and two objects of identical size (e.g. two triangles) are fixated and fused into one image (Fig. 5.2a). Whilst maintaining the distance from the fixating eyes constant, the difference between the two objects is varied such that the eyes (assisted by the fusion tendency) are subjected to convergence angles of 7–60°, whilst accommodation remains constant. With increasing convergence, the perceived size of the unchanging objects decreases rapidly ('convergence microsia'). On the basis of curves derived from such tests, the object-size is then varied in such a way that the perceived size remains constant throughout the entire convergence range. From the angle of convergence and the appropriate object-size at each point, the participation of the convergence mechanism in the correction of the retinal image can be calculated as a reciprocal function: curve *c* in Fig. 5.1

Accommodation. Two identical objects are once again viewed with crossed visual axes, and the image is fused. In this case, however, the angle of convergence is maintained constant whilst the objects are moved from 50 cm to 8 cm away from the fixating eyes (Fig. 5.2b). As before, curves derived from initial tests are used to vary the objective size of the objects such that the *perceived* size remains constant throughout the entire range of distance. The data obtained

---

[1] All of the curves were derived from experiments on the author himself. Preliminary tests on other experimental subjects indicate that there may be small individual differences.

permit calculation of the part played by the accommodation mechanism in the correction of the retinal image: curve *d* in Fig. 5.1. This part is roughly equivalent to that played by convergence alone.

Curves *c* and *d* lie in a roughly intermediate position between *a* and *b*. If the distances at each point between *a* and *c*, and *a* and *d* are added, one obtains curve *e*, which virtually coincides with *b*. The correspondence of curve *b* with curve *e* is better than appears at first sight. At no point does the sum of the image reduction exerted by the two separate mechanisms differ by more than 8 per cent from the required value. In other words: The growing retinal

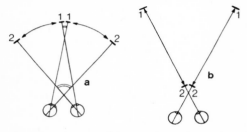

Figure 5.2. Diagrams of the experimental procedures. In *a*, only convergence is varied; in *b*, only accommodation. The width of the objects, which are perceived as one fused image and as of constant size, is indicated for the two extreme positions (1 and 2) in each case.

image produced by an approaching object is centrally reduced by the activity of the convergence mechanism or the accommodation mechanism alone such that constancy is maintained to exactly half the required extent. If both mechanisms are functioning simultaneously, the correcting effect is added.

It is in this way that convergence and accommodation operate in the *simplest* fashion (that is, equally and through addition) so that we are able to see things as big as they 'really' are.

6 Is the influence of accommodation on visually perceived object-size a 'reflex' process? *1955*

Size-constancy of visually perceived objects is attained through additive interaction of the mechanisms for convergence and accommodation. To approximately equal extents, these two mechanisms reduce the growing retinal image of an approaching object as it progresses to the level of perception (and, conversely, magnify the shrinking image of a departing object). The question arises as to whether this central correction process is set in action by receptors in the eye muscles and the muscular apparatus of the lens which are stimulated by movements of convergence and accommodation, respectively. In the terms of the generally current interpretation of nervous functional systems, this question would automatically be

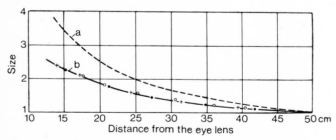

Figure 6.1. *a*: alteration of the size of the retinal image of an object according to its distance from the eye. *b*: alteration of the visually perceived size of an object under various experimental conditions with constant convergence (20°). (The size at a distance of 50 cm is taken as unity.)

given a positive answer. However, this can be proved to be erroneous in the case of accommodation.

Two objects (triangles) are fixated with the visual axes crossed, with the angle of convergence maintained constant, and the images are fused. The objects are then brought closer to the eye, at the same time altering their size such that they are perceived to be of constant dimensions. From the reciprocal values of the size-data, one can obtain the curve *b* of Fig. 6.1. This curve shows the extent to which an object is perceived to become bigger as it approaches the eyes, whilst curve *a* of Fig. 6.1 indicates the simultaneous increase in size of the image on the retina (see pp. 185–7). If the peripheral accommodation apparatus of one eye is now anaesthetized with atropine, so that this eye remains consistently adapted for distant vision, the following variants of the above experiment can be carried out:

1. The paralysed eye fixates a distant, stationary light spot (in order to maintain the angle of convergence constant), whilst with the aid of the intact eye an object approaching from a distance of 50 cm is varied so that size-constancy is exactly maintained. From the required alterations of the size of the object, it is possible to obtain a curve ●−● in Fig. 6.1) which coincides with that for *two* intact eyes undergoing accommodation. Thus, the accommodation mechanism of the intact eye remains unaffected by the suspension of operation of the eye on the other side, and its reducing function retains its full strength. The two eyes, along with their associated central mechanisms, carry out their size computations independently.

2. The accommodation-paralysed eye is presented at 4–5 m distance with an object (triangle) which is distinctly seen, and which is of such a size that its image exactly fits with that of an object seen at a distance of 50 cm with the intact eye (Fig. 6.2). Then, by approaching the latter object to the intact eye as in experiment 1, the accommodation command is set in action. This only applies to the intact eye, however, while the paralysed eye continues to receive a distinct image of the distant object. It emerges that the distant object viewed with the anaesthetized eye, despite the constant size of the retinal image, does not retain its perceived size, but becomes considerably smaller ('accommodation micropsia'). Fig. 6.3 shows the extent to which this object must be *enlarged* so that it is perceived to be of constant size, and so that it fuses

to form a clear image with the object seen by the intact eye at any distance. If one then extracts from Fig. 6.3 the corresponding data indicating how much the retinal image of the paralysed eye

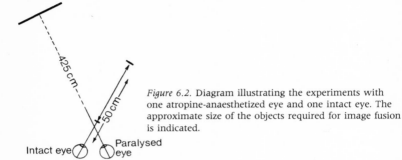

Figure 6.2. Diagram illustrating the experiments with one atropine-anaesthetized eye and one intact eye. The approximate size of the objects required for image fusion is indicated.

is perceived to be smaller with increasing 'voluntary' adaptation for close-up vision, the values (o−o) once again lie on curve *b* of Fig. 6.1. Therefore, this correcting process takes place in the *same quantitative manner* in both the intact and the paralysed peripheral mechanisms. But since exclusion of the accommodation movement must rule out any reflex which the latter might elicit, we cannot in this case be concerned with a stimulus-dependent process of this kind. Only the accommodation command itself can be involved as the cause of central image reduction.

3. If the paralysed eye is covered to leave a small aperture of about 0·5 mm diameter, distinct vision is possible with objects quite close to the eye (though there is a loss in light intensity). If the

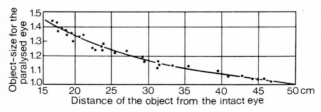

Figure 6.3. Objective increase in size of an object fixated with an atropine-treated eye and visually perceived to retain a constant size (distance 425 cm), with simultaneous accommodation of the intact eye to an object approaching from a distance of 50 cm (cf. Fig. 6.2).

experiment is then repeated such that two objects approach *both* eyes from a distance of 50 cm, they only fuse into one image and appear to be of constant size when they are maintained equal in size and simultaneously reduced to the same extent as would be necessary for two *intact* eyes (+ − +; Fig. 6.1). This provides further confirmation for the conclusion drawn from experiment 2.

Thus, the constancy function of the accommodation mechanism is a process of a purely *central* kind. There is some indication that the same also applies to the constancy function of the convergence mechanism.

7 Active functions of human visual perception *1957*

DEDICATED TO KARL VON FRISCH ON HIS
70TH BIRTHDAY

It is the task of the senses to provide the organism with information about its environment with sufficient exactitude to permit it to live within that framework. This banal sentence incorporates problems which have only been resolved to a very slight extent and which would still be present even if the activity of the sensory mechanisms had been exhaustively studied. It is not sufficient for the sensory organs to pick up 'stimuli' with maximum exactitude, and for the organism to 'respond' to these in a particular fashion with 'reflexes' or with acquired behaviour. In order to exist, the organism must recognize the most important *objects* as such under an extremely wide range of conditions, and this recognition must be independent of the 'stimuli' emitted by the objects at any instant and the number of these 'stimuli' reaching the organism. It is impermissible that a lightly-tinted object should be seen as a dark one because little light happens to fall upon it at a given time, that a loud sound should be heard as a weak one because it comes from far away, that an immobile object should be seen as moving because the eye is passing across it (such that the image wanders over the retina), that a big predator should appear to be a small prey organism because it is a long way away, or that a white object should appear red beneath the evening sun and green underneath foliage. The greater the number of distinctions made by the sensory organs, the more numerous the possibilities of error which arise.

192

Experience shows that organisms – including man – do not generally fall prey to these errors. In addition, we know for many animal species that they behave in a correct manner from the very outset of their active life, such that learning cannot play a considerable part in this respect. For human beings, this question cannot be so easily settled, since many of our life functions mature slowly. We know that some aspects of our perception, too, can be modified by learning – and many people have concluded from this that the entire attribute is presumably based on experience alone. However, it is emerging with increasing clarity that we also possess an innate apparatus which performs these functions. Such apparatus is, to a certain degree, plastic and modifiable, as must necessarily be the case since a level of precision is required which far exceeds that possessed by morphological structures emerging during embryonic development. Thus, the 'fine adjustment' is provided in the course of juvenile development (ontogeny). Nevertheless, it appears that experience, learning and (above all) insight have amazingly little influence upon the processes to be discussed here.

To restrict ourselves to human beings, how is it possible that a man is able to distinguish between a weak sound from a source close at hand and a loud, distant source of sound? In other words, how is one able to say, independently of the intensity with which the eardrum is set in motion: 'This noise or this sound "is" loud or faint'? In the field of vision there is a whole range of such achievements. The best known concerns the perception of brightness. An object is seen with a particular *inherent* brightness which is largely independent of the light intensity transmitted by it. The extent of this independence only becomes uncomfortably obvious to a photographer who takes pictures without an objectively operating lightmeter. During the bright midday period one overexposes, whilst in the evening one underexposes.

Such functions are referred to as *constancy* functions in psychology, and one talks in terms of loudness constancy, brightness constancy and so on. At first sight, it seems as though the organism must possess the ability to bypass what is provided by the physical operation of the sensory apparatus and to directly reach what is really interesting – the *objects* themselves. But this is not the case. There is no bypassing of physical factors; instead, the latter are exploited down to their finest details in a refined manner to permit manifold correction. With *loudness constancy* it is presumably the change in intensity relationship between the overtones with distance which, above all, informs us of the distance of the sound source, and the loudness of the source is deduced from this relationship. This is achieved by an apparatus in which there is no conscious participation, and it can consequently easily be led astray. If we encounter sounds whose tonal spectrum departs from the usual pattern in a manner otherwise found with distant sounds, we fall prey to an illusion: we believe that the producer of this sound is farther away and that the sound source is louder than it actually is. This is what happens to the call of the cuckoo, which seems to ring out far across the countryside and then abruptly shrinks when the sound-producer is discovered unexpectedly close at hand. The oft-bewondered 'carrying capacity' of a pure violin note is based on a similar principle: close at hand it is subdued, whilst it literally becomes *bigger* when farther away.

With *brightness constancy*, the situation is – in principle – different but it is still extremely simple. The apparatus which operates in this case and which (in vertebrates) determines the diameter of the visual aperture, the depth position of the visual elements in the retina, their differential shielding through migration of pigments, the equilibrium state of light-sensitive substances, and (finally) the sensitivity of visual centres, is simply dependent upon the average brightness of the entire visual field encompassed at any time. The fundamental principle is to preserve, as far as possible, a 'comfortable' brightness – independently of the momentary light input – such that objects are perceived with clear contrast, as in a 'correctly' exposed photograph. The functional limitations of this vital apparatus are well known to the nocturnal car driver, whose crepuscular visual sphere immediately disappears into darkness when his eye is struck by the headlights of an oncoming car.

It is not quite so easy to understand a further constancy function which consists in judging with fair accuracy the *position of objects relative to the ground*, independently of the position which the human observer happens to adopt (and thus independently of the orientation of the image in the eye). We carry within us (of course, at the subconscious level once again) a coordinate system which is related to gravity and in which the visually perceived matter is traced according to its orientation to gravity. This function is made possible by the activity of a sensory apparatus located in the labyrinth – the statolith apparatus. The latter consists essentially of small stones resting on sensory hairs which are bent to varying extents and in different directions by the stones according to the position of the head. This bending produces a stimulus configuration which transmits impulses to a low-level postural centre. The latter, in its turn, literally *calculates* from these impulses the momentary position of the head and thus makes available the data for continuous correction of the position of the image on the retina.

Such an apparatus can also be deliberately misled. If I sit in an enclosed centrifuge chamber which is revolving with a constant velocity, what I perceive as 'below' is the direction of the resultant of gravity and the centrifugal force, since it is in this direction that the pressure of the stones in the labyrinth is applied. Naturally, as soon as I incline my head to one side (departing from the resultant of the two forces), the bending of the sensory hairs will be much greater than under normal conditions. In other words, for a particular angular departure the postural centre will calculate a far greater angle. The result is that I unavoidably see things turning with my head in the centrifuge, since they pass over the retina to a lesser extent than that calculated – and compensated for – by the postural centre. To put it another way: the subconscious coordinate system over-reaches!

There is an equally impressive perceptual effect when a quite weak electric current is sent in alternation across my head from one mastoid bone to the other, a current which cannot be felt as such but which is sufficiently strong to stimulate the postural centre rhythmically. The outcome is that there is a rhythmic oscillation of the observed environment about the axis from the forehead to the back of the head. Once again, this is because my subconscious reference system relative to gravity is induced by the stimulation to oscillate rhythmically.

As can be seen, the mechanism of this constancy function (which can be referred to as *vertical constancy*) is fundamentally different from that involved in brightness constancy. In this case, two sense organs are operating *in opposition*. When I tilt my head to one side, the image of the environment which I am observing moves in the opposite direction across the retina. Without any correction, I would necessarily be led to believe that the environment had moved. But the statolith apparatus is simultaneously activated and readjusts my gravity-oriented subconscious orienting system such that the visually induced error is roughly compensated.

There is, in fact, a much more surprising type of constancy function in which the data necessary for correction of disturbance, or 'illusion', are derivable neither from the receiving sense organ (as with loudness and brightness constancy), nor from any other sense organ operating in opposition (as with vertical constancy). This is simply because the sensory organ concerned does not provide such data and because there is no available counteracting organ. It is here that actual *active functions* of the optical perception system begin to operate, and these shall be the main object of our attention. So far, we have been able to talk in terms of refined central evaluation of the stimulus current passing into the nervous system – the *afference* – and to regard its correction as determined by external stimuli. However, this is no longer possible in the following case. Here, there is emergence of a new, corrective force which originates from the nervous system itself and which can in fact be expressed in *spontaneously produced perception*. I shall attempt to demonstrate that a large number of so-called 'optical illusions', some of which have been known for some time and are regarded as curiosities, as disruptions in the field of human perception, are in reality the opposite of disruptions – they are the means by which peripherally produced disruptions are compensated and through which constancy functions are therefore ensured. The two most important constancy functions for which such evidence is to be presented are *directional constancy* and the *size constancy* of viewed objects.[1]

By *directional constancy* I mean the fact that we locate objects in our spatial environment in the 'correct' direction; that is, that we perceive immobile objects as immobile and moving things as mov-

[1] The concept elaborated in the following account was first indicated by experiments with animals (von Holst and Mittelstaedt, 1950). Although this concept was apparently not formulated in the literature prior to 1951, it is so plausible that it was quite probably apparent to earlier authors, who were perhaps prevented from taking it seriously or formulating it clearly by the prevailing accepted theories. In the physiological and psychological literature of recent years, however, this concept seems to be gaining ground.

ing, independently of the movements which we ourselves perform (i.e. with the body, the head and – above all – the eyes). For example, when I stand upon a railway platform and keep my eyes fixed straight ahead on a train which is starting to move, the image of the train moves across my retina in exactly the same way as when the train is standing still and my gaze is actively moved past. In

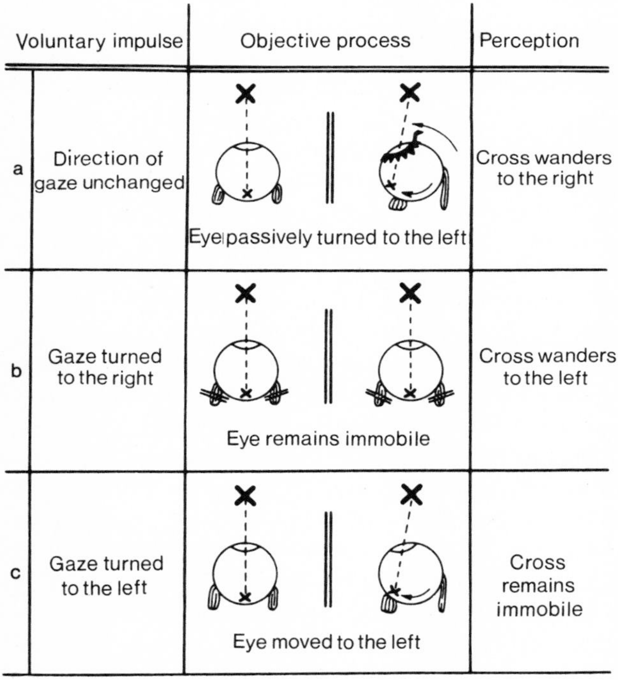

	Voluntary impulse	Objective process	Perception
a	Direction of gaze unchanged	Eye passively turned to the left	Cross wanders to the right
b	Gaze turned to the right	Eye remains immobile	Cross wanders to the left
c	Gaze turned to the left	Eye moved to the left	Cross remains immobile

Figure 7.1.

the first case, however, I see that the train is moving off, whilst in the second case I see it as standing still. How is this vital distinction achieved?

Let us perform a simple experiment. On a blackboard placed in front of me, I draw a cross at eye level, and two hand widths to the left I add a circle. I now look at the cross with both eyes and then mechanically fix the position of the right eye (previously rendered insensitive to touch) with a small enveloping clamp. I then turn my gaze to the circle on the left. My left, freely movable eye is then directed to the circle, whilst the right (mechanically fixed) eye cannot follow the desired movement. Surprisingly, I perceive that at the instant when I turn my gaze to the circle, the cross leaps into it! If we dissect this initial observation to some extent, to expose its elements, we can immediately see how it arises.

We can consider the behaviour of just the one eye, keeping the other closed for the sake of simplicity. Three experiments are performed one after the other with the open eye (Fig. 7.1). The eye is first directed straight ahead towards an object of some kind, for example a cross (7.1a). The enveloping clamp already mentioned above is then attached, and the eye is mechanically (i.e. passively) turned to the left. When this is done, I perceive the observed cross to wander to the right. The second experiment (7.1b) involves use of the clamp for fixation of the position of the eye, which is once again directed towards the cross, followed by the *intention* to move the gaze to the left. This movement is not performed by the eye since it has been fixed; nevertheless, I perceive a movement once again – the cross wanders to the left! This experiment is particularly important, since it shows that perception of movement can occur even when the image of the retina does not actually move at all. Thus, such perception cannot originate from any peripheral stimulation, from some kind of afference.[1] Finally, as a third experiment (Fig. 7.1c), the two experiments outlined above are combined. The clamp is fitted to the eye whilst it is directed straight ahead and observing the cross, the intention to turn the gaze to the left is then produced and, at the same time, the eye is moved in the same direction using the clamp. If this experiment is adeptly performed, the result is: the observed cross remains immobile. What we have just done is nothing other than what normally happens with free movement of the eye. There is an intention to turn the gaze to the left, the eye performs the turning movement, and the observed environment remains immobile. This is, in sum, the phenomenon of directional constancy. But, at the same time, we already have the key to explanation of the problem. Directional constancy is the product of the emergence of two mutually opposed optical illusions. One

[1] This second experiment can also be performed without mechanical fixation of the eye, if the musculature for moving the eye is temporarily paralysed (Kornmüller, 1947). The result is then exactly the same. This eliminates the possible explanation that sensory cells in the musculature conceivably stimulated by the resultant musculature tension could be the cause of the described perception of movement.

(experiment 7.1a) is the result of the movement of the image across my retina; the other (experiment 7.1b) is the result of my intention – my voluntary command – to turn my gaze. When the two occur simultaneously, they cancel one another out completely (Fig. 7.1c) and the result is that we see the observed environment as immobile, despite the active movement of our eyes.

This simple functional relationship can be represented with a schema (Fig. 7.2). Only my voluntary impulse (e.g. the intention to move the eye to the left) and what I consequently perceive is conscious. What happens simultaneously at the subconscious level can be briefly described as follows. A higher centre transmits a 'command' for movement of the eye, and this leads to corresponding rotation of the eyeball through impulses sent to the musculature. This rota-

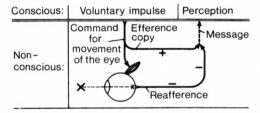

Figure 7.2.

tion produces migration of the image across the retina. This peripheral process is the necessary 'reflection' of my eye movement. Since the expression 'reflex' is otherwise inappropriate, we can refer to this as a *reafference*. If this reafference occurred in isolation, I would necessarily see the environment leaping around with any movement of my gaze. But experiment 7.1b demonstrates that my voluntary impulse not only elicits a motor command (referred to as an *efference* in physiological terms) but also gives rise to a branch which itself produces a perceptual process. The latter is of the same kind as the reafference, only it has an inverted directional sign. This branch, which can be referred to as the *efference copy,* usually meets up with the reafference, and the two then cancel one another out. Objects remain in their places because no message is sent up to the perceptual level. It is only when the efference copy is eliminated because there is no command (Fig. 7.1a), or when the afference is cut out because there is no eye movement (Fig. 7.1b), that the observed picture is made to leap around (in opposite directions in these two cases).

This schema does not incorporate a theory of any kind, but simply a description of the factual situation which we have encountered. With its aid, we can now predict certain things which could not previously be explained. Let us assume, for example, that I see a seagull flying across a blue sky. This can happen in two ways: either I keep my eyes immobile and pointing in a particular direction, in which case I see the gull move across my visual field; or I follow the flying gull with my gaze, turning my eyes to keep up with it. In both cases, I perceive the direction and approximate speed of the flight, although in the second case nothing is moving across my

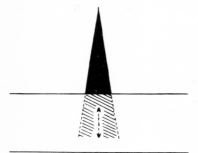

Figure 7.3.

retina. It is now easy to see the explanation for this. In the first case, there is no motor command and perception of the movement is the result of migration of the gull's image across the retina – the afference. In the second case, no such image migration occurs, and perception of the movement is a result of the motor command, i.e. the efference copy, which is not cancelled out in this case. In the latter case, the gull is correctly seen to be flying forwards because we are equipped with a quite specific (wrongly labelled) 'illusion' coupled to the voluntary impulse!

Let us now turn to another phenomenon – that of *size constancy* – where the relationships are basically the same, but where there are more factors involved. By size constancy we mean the fact that a given object (even an unfamiliar one) is not seen roughly as it is portrayed on the retina, sometimes large and sometimes small, but is perceived at approximately its 'correct size' whatever its distance from the eyes may be. The achievement involved can be quantitatively measured against an optically homogeneous background by moving a triangle towards and away from the eyes and at the same time reducing or increasing the size of the triangle with a mask (Fig. 7.3) such that it is perceived to remain exactly the same size. For myself, I obtained the curve shown in Fig. 7.4. (Other experimental subjects exhibit insignificant differences which need not concern us here.) Thus, in the range from 50 to 10 cm away (in the 'grasping range') I perceive a triangle of this kind to retain exactly the same size if it is reduced by about one-fifth during the approach. If I wished to keep the size of the image on the retina constant, I would have to reduce the triangle by about nine-tenths in size.

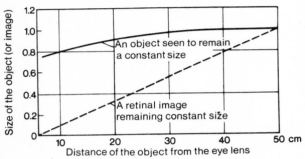

Figure 7.4.

The same result can also be represented in a different fashion by indicating the quantitative perceived increase in size of an objectively constant triangle occurring when the latter is approached from 50 to 10 cm away (Fig. 7.5b). The resulting curve is almost horizontal, whereas the image of the object in the eye is at the same time rapidly growing in size (Fig. 7.5a). Thus, the constancy function is quite considerable within the 'grasping range', which is so particularly vital to us. How is this function achieved?

Drawing on what we have previously discovered about directional constancy, we can presume that in this case, too, there is participation of the voluntary command, which could be coupled to active motor processes. In fact, we know of two active motor processes involved in adjustment of the eyes for distant and close vision. These are: *convergence* (i.e. the alteration of the angle between the visual axes of the two eyes) and *accommodation* (i.e. focussing of the lens).

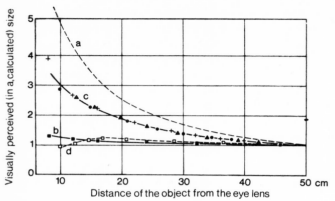

Figure 7.5.

a -- size of the retinal image
b ■ visually perceived object size with binocular fixation
c ● visually perceived object size with accommodation maintained constant
c + visually perceived object size with constant convergence (40°)
c ▲ visually perceived object size with constant convergence and paralysed accommodation
d □ twice the distance of curve c from curve a

Convergence and accommodation are generally combined, but we can experimentally separate them for independent testing.

Instead of directing our gaze at *one* object, as is usual, we can (after some practice) cross our visual axes so as to look at one triangle with one eye and a second adjacent triangle of the same size with the other eye, thereby leading to their combination as one image in perception. It is now quite easy to vary the distance

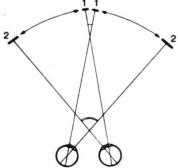

Figure 7.6.

202

between these two equally large triangles without altering their distance from the observing eyes; that is, without any involvement of accommodation (Fig. 7.6). The so-called 'fusion compulsion' ensures that both eyes follow any such movement. Thus, when the triangles are separated (i.e. moved from 1 to 2 in Fig. 7.6), the two visual axes are increasingly crossed and the eyes are increasingly converged' *as if* one were observing an approaching object. When the two triangles are kept the same size, the result of this experiment is quite amazing. The greater the convergence forced upon the eyes, the smaller the observed triangle becomes. One can also arrange the experiment such that the two displayed triangles are objectively increased in size such that, at any angle of convergence, perception

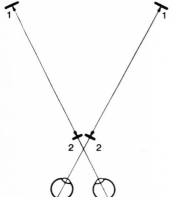

Figure 7.7.

reports a triangle of the same size. From the triangle sizes which are necessitated, it is easy to calculate the extent to which, with increasing convergence, an image of constant size on the retina is reduced through the process of convergence itself, i.e. by the voluntary command. The curve in Fig. 7.5c shows the result. Thus, in this case we are concerned solely with convergence, since accommodation remained unchanged.

As a next step, we can similarly investigate accommodation in isolation. Once again, we use both eyes to look at two triangles of equal size with the visual axes crossed. But this time the angle of convergence is maintained constant and the two triangles are moved towards and away from their respective eyes along two guide-rails (Fig. 7.7). In contrast to the normal case, approach of the triangle (from 1 to 2) causes it to be seen as much larger. The objective size of the displayed triangles is accordingly altered such that the triangle is always seen to be the same size. From the measurements taken in this way, it is once again easy to calculate the diminution effect that close-up focussing of the eye has upon the retinal image. Once more, Fig. 7.5c demonstrates the curve obtained. The course of the curve is (for myself) the same as that for the diminution curve for pure convergence. Both curves lie almost exactly halfway between the normal curve for size constancy (7.5b) and the curve given by the increase in size of the retinal image (7.5a) accompanying approach of the object. This means: the voluntary act of increasing convergence and the voluntary act for increasing close-up focussing of the lens are each responsible for half of the correction of the retinal image. If the two effects are added together, we arrive at the curve 7.5d, which is quite satisfactorily in agreement with the initial curve for size constancy.

Thus, the assumption that convergence and accommodation have something to do with size constancy has been fully supported. The question remains open as to whether the voluntary command itself is the cause of the correction or whether sensory elements which are stimulated through the motor process are perhaps responsible for correction, as we have in fact seen with vertical constancy. Such an assumption is scarcely likely for the convergence movement in view of the facts which we have already discussed with respect to directional constancy. In the case of accommodation movements, the question can be tested experimentally. It is quite simple to paralyse accommodation for some time by applying a few drops of atropine to the eye. This poison paralyses the circular musculature which normally produces pronounced bulging of the lens with close-up adjustment of the eye. As a result, the lens remains stretched and flat (i.e. adjusted for distant vision) between the elastic suspensory fibres even when the subject wishes to look at objects close by. If this is attempted, there is emergence of a persistent sensory illusion referred to as *micropsia*. For example, if I close my left eye and look at the far corner of a room with my atropine-treated right eye, my perception is normal. But if I now attempt to fixate the tip of my finger just in front of my eye, the attempt fails and the entire room (which I continue to see in focus) shrinks to the proportions of a doll's house. In this case, as in the experiment 2 with inhibited eye movements (Fig. 7.1b) described above, *absolutely nothing happens at the periphery*. The image on the retina remains the same, the eye lens does not move (such that there is no possibility of stimulation of any local sensory cells) – and nevertheless I perceive the observed environment as shrinking in size! Consequently we must conclude that in this case, too, the sensory illusion represents *a direct result of my voluntary command*.

It is also possible to take exact quantitative measurements of the extent of this microptic reduction. The result of such measurement reads as follows. Reduction produced by accommodation when the apparatus is functional and the reduction produced with atropine

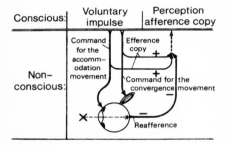

Figure 7.8.

paralysis are exactly equivalent. This means that the voluntary command (more exactly, the efference copy of this command, to use our terminology) is the sole cause of the perceptual effect described.

In this case, too, an overall summary leads to a functional schema corresponding to that for directional constancy, except that here we must divide the voluntary impulse for close-up adjustment into two different command channels (Fig. 7.8). One channel leads to the convergence movement, and its efference copy produces a perceptual illusion of a particular magnitude. The second leads to the accommodation movement, and its efference copy produces a perceptual illusion of roughly the same magnitude and in the same direction. The approach of an object to the eye produces enlargement of the image on the retina and this itself leads to a perceptual illusion; the latter is equal in magnitude to the sum of the two efference copies and operates in the opposite direction. If the three 'impulse streams' combine, they are mutually cancelled to produce a zero effect. Thus, when the apparatus is intact, my impression is that an approaching object remains virtually constant in size.

It is remarkable that the size of a viewed object is *directly* perceived; it is unnecessary for the object to be first seen at a given distance. In the experiments with convergent visual axes described above, the size is always perceived with great exactitude, whilst estimation of the distance is extremely variable and unreliable. This is understandable, since the two mechanisms involved (convergence and accommodation) are not cooperating in the normal manner.[1] The established theory, according to which the size of an object is determined indirectly through its estimated or 'imagined' distance, is therefore invalid. The apparatus in my lower centre, which compares my voluntary and accommodation impulses against the dimensions of the retinal image in order to calculate the size of an object, does indeed utilize data which also play a part in the estimation of distance; but these two non-conscious calculating procedures are independent of one another. For this reason, these procedures probably take place at different sites.

The functional schema in Fig. 7.8 is nothing more than a description of the factual relationships outlined. But, apart from elucidating size constancy, it also explains other phenomena, only one of which need be mentioned here. If we fixate a luminescent cross for some time in a dimly illuminated room, the image 'engraved' upon the retina persists for a number of seconds as an 'after-image'. With this after-image in the background of the eye, we can look (for example) at the far wall or at a nearby sheet of paper – wherever we look, it will be there. But the size of the cross varies markedly: on the distant wall it is extremely large, whilst on the sheet of paper it becomes tinier and tinier as the paper is brought towards the eye. At first sight, this is amazing, since the image on the retina must, of course, retain its original size. However, with respect to our functional schema this result is to be expected. If I adjust my eyes for distant or close-up vision with a retinal image which remains the same size, I naturally produce a voluntary command which directly passes to the conscious level as a perceptual effect, since there is no opposing afference to cancel the process on its way. Thus, the factor which makes the perceived cross become large or small in

---

[1] Calculation of distance is not additive, but apparently follows the principle of double-checking'. One *datum* should confirm the other, and in the experimental situation there is a consequent insoluble contradiction.

this case is, once again, nothing other than an active function of the perceptual apparatus.

So far, we have only given detailed attention to the important phenomenon of size constancy in the range of the *grasping zone*, within which accommodation and convergence play a large part for purely geometric reasons. The greater the distance of the objects from the observer becomes, the lesser the exactitude with which these two mechanisms can perform their functions. However, in the space towards which we are moving at any given time, in which the above two mechanisms have already suffered attenuation, quite different

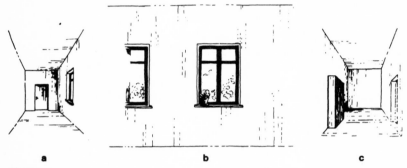

a   **b**   **c**

Figure 7.9.

correcting functions predominate. These are remarkable enough to justify closer investigation.

Once again, we can set out from a simple observation. Assume that I am standing in the middle of a long corridor and that I turn to face the far end. What I perceive corresponds roughly to the sketch in Fig. 7.9a. I then turn through 90° and look at the side wall of the corridor which is now facing me, and my perception reports roughly the picture in Fig. 7.9b. On turning a further 90°, I come to face the other end of the corridor, and I see the picture shown in Fig. 7.9c. These three sketches placed side by side present us with a particular problem. Whatever direction one chooses to look along the corridor, its contours (such as the angles at the floor and the ceiling) are seen as *straight* lines running in quite *different* directions, such that they would have to form either a bend or a curve if they were to be joined up. How is it that I do not perceive such a bend or curve? If one thinks of the described movement through 180° as taking place slowly and then attempts to draw a reconstruction of the image wandering across the retina,[1] one in fact obtains the sketch shown in Fig. 7.10, in which the straight horizontal lines of the room appear as curved in a particular fashion. How is it, then, that I do not perceive the corridor as such a distorted phenomenon, but see it as it 'really' is? The answer to this question cannot be given in one sentence,

Figure 7.10.

because in this case it seems to me that several (at least three) correcting mechanisms of quite different kinds are cooperating hand in hand, all without conscious participation.

(1) In the first place, it should be noted that so-called 'perspective' vision of depth (Fig. 7.9a, c) apparently represents just *one* visual possibility which – without normally coming to our notice – can occasionally be exchanged against *other*, more appropriate possibilities, which have already been mentioned. When I am looking along an avenue with a view to drawing it, I can easily trace perspective lines of the kind illustrated in the sketch of Fig. 7.9a. But, apart from this ability, I can at any time make use of the mechanisms already described to see the correct size of the individual trees in the avenue. When I am walking casually along the avenue, the latter kind of visualization is predominant. This is also the pattern of visualization found in children. Usually, their drawings depict distant and nearby houses and people as being the same size. This is not because

[1] More exactly: across an imaginary vertical line on the retina.

they 'know' that the sizes are the same, but because they *see* them as being the same size. It is only later on, with habituation to perspective viewing and with continuous exposure to photographs, that the other pattern of visualization gradually becomes predominant. This pattern is one which is also predominant among many contemporary painters; but this was by no means the case in the paintings of past centuries.[1] Accordingly, perspective vision is doubtless a late acquisition, both phylogenetically and ontogenetically, while the

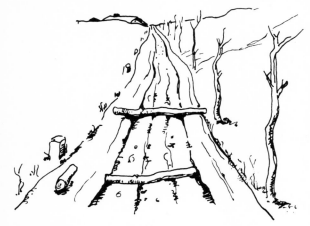

Figure 7.11.

mechanisms of size constancy are an age-old property of visually oriented animals.

Closer examination shows, however, that even with a normal perspective image our perception carries out a remarkable form of correction which is known as an 'optical illusion'. It is necessary to make a close quantitative study of this effect. In the sketch (Fig. 7.11) of a pathway with two tree trunks lying on the ground, the one in the foreground seems much shorter than the one in the background, despite the fact that they are exactly the same length when measured in millimetres on the illustration. If a series of such sketches is prepared, in each of which the length relationship of these two trunks varies by about 1–2 per cent, and a large number of observers are asked where they would expect the trunks to have

[1] It first appeared in Leonardo da Vinci's time, when the so-called 'Laws of Perspective' were discovered.

the same length in millimetres on the paper, an objective length relationship of 1 : 0·85 is obtained (40 experimental subjects; mean deviation 0·051 per cent). With such an example one is at first naturally inclined to believe that this so-called 'illusion of judgement' is based upon experience – on learning. But this is only true to a limited extent, if at all, as is illustrated by two facts. Firstly, one can present the same picture to initially unaccustomed experimental subjects as an inverted image (describing it as a sketch of a histological preparation or such like). In this case, the two rods

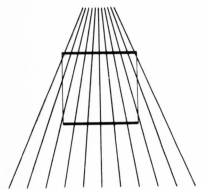

Figure 7.12.

or trunks are not seen as having the same length; they only appear to be equally long with a length ratio of 1 : 0·846! Thus, insight into the subject represented by the sketch obviously has no influence whatsoever on estimation of length. Instead, the latter involves a perceptual correction exclusively evoked by certain linear features. The second argument comes from a so-called 'geometrical-optical illusion' in which the corresponding linear features are similarly presented in pure form, as a model (Fig. 7.12). The extent of this illusion, when quantitatively tested on the same experimental subjects, is such that the length ratio between the lower and upper bars in the square (Fig. 7.12) must be 1 : 0·893 if the two are to be perceived as having the same length (mean deviation: 0.03 per cent).[1]

[1] The correction performed in this case is, therefore, less than that executed for the tree trunks lying across the pathway (statistically significant). One might therefore think that there is involvement of an empirical factor if the result of inversion of the picture did not speak against this. The difference probably derives, however, from the fact that the sketch in Fig. 7.11 is more plastic and more impressive.

210

To take a second example: Fig. 7.13 shows an axe with its cutting edge directed towards the observer alongside a partially open book balanced on one end. What one should do in this case is compare

Figure 7.13.

the length of the cutting edge and that of the binding of the book. The two are perceived as having the same length when their actual length ratio is 1:0·825 (40 experimental subjects; mean deviation 0·033 per cent). This figure also incorporates a linear schema, a model, which provokes a perceptual size correction in its own right. This is the well-known geometrical-optical 'illusion' shown in Fig. 7.14. Here, the two vertical lines must have a length ratio of 1:0·832 (mean deviation 0·043 per cent) in order to appear equal in length.

Figure 7.14.

211

One therefore obtains the following factual conclusion from this investigation. We are *unable* to perceive the objective proportions of even a perspective drawing projected onto the retina because an optical correcting mechanism alters these proportions[1] in the direction of size constancy in a non-conscious fashion – although convergence and accommodation play no part (and would, at the most, inhibit this process).

(2) A second, quite different, correcting mechanism is demonstrated by the following experiment. Assume that I am sitting in a dark room, looking at a small spot of light localized on the wall directly in front of me. At the same time, another light-spot slowly oscillates to and fro across the wall at the same height. If I fixate the localized light-spot, I can also clearly see the oscillating spot passing backwards and forwards along a horizontal line. The two light-spots are then moved upwards by six to nine feet, and – with my head held straight – I turn my eyes upwards to fixate the immobile light-spot near the ceiling, whilst the other light-spot oscillates to and fro at the same height. In this case, too, I perceive the movement of the oscillating point as a straight line. However, as soon as the immobile light-spot is extinguished and I fixate the oscillating spot, following its movement with my eyes, it no longer moves along a straight line but along a curve corresponding to the upper margin of the corridor in Fig. 7.10. If the light-spot oscillates below my eye level, the curve is inverted and corresponds to the lower margin of the corridor wall in Fig. 7.10. It only moves in a straight line at eye level. This impressive observation demonstrates that there must be a correcting mechanism which is able to 'bend' the curved lines running across the retina in Fig. 7.10 to give straight lines, but that this only happens when the eye itself is at rest and is not forced to move uniformly.

We cannot yet state the basis for this correcting mechanism. Nevertheless, one thing is important in this respect: our *natural* eye movements *never* occur in a smooth, uniform fashion. We do indeed

---

[1] Anyone acquainted with 'dummy'-experiments on animals will immediately appreciate this interpretation of various geometrical-optical 'illusions', and the author has long been aware of this explanatory basis. A survey of the literature on human psychology showed this to be presented (at least in outline form) by O. Klemm (1919), who speaks of 'perspective auxiliary interpretations' in relation to such illusions, and more recently (without reference to Klemm, and without quantitative data) by R. Tausch (1954).

believe that our eyes *glide* over the lines of a book we are reading, or across a landscape; but this is only an inherent feature of our perception. The eye movement itself occurs as *leaps*, with three to five such leaps spanning the line of a book, and with fairly large leaps covering an observed landscape. Between these leaps, the eye remains still for an instant – and it is only during this moment of rest that the picture is actually received. During the leaps (which are, in fact, predominantly accompanied by eyelid movements), there is apparently no perception. Thus, the eye *strides* rather than glides. This striding is, as we can now see, necessary so that the distortions arising on the retina for geometrical reasons can be corrected. Therefore, when we stand in the corridor and turn our gaze from one end to the other, ultimately producing an objectively accurate overall perceptual image of the corridor, this is not at all the result of lining up or 'smearing' of component images produced on the retina. Instead, the whole phenomenon represents a *higher integrative process* for which mere reception on the retina simply provides raw material that is badly in need of refinement.

(3) There is also a third mechanism, again quite different, which I believe to be significant in this context. This is the well-known physiological phenomenon of *'Listing's eye rolling'*. As long as one is looking straight ahead, the eye is simply turned around a vertical axis whenever the gaze wanders to the right or to the left. But if the eye is, for example, directed obliquely upwards and follows the upper margin of the room, it no longer rotates around a vertical axis. Instead, it rotates around an axis which is inclined backwards at the same angle from the vertical as that between the upward axis of the gaze and the horizontal. This means, however, that the eye rolls, and it rolls in such a manner that it approximately follows the perspective sweep of the upper margin of the room, as shown in Fig. 7.10. Thus, when the eye is turned to the left or the right, the fragment of the margin envisaged at any time lies roughly on the *same* area of the retina and accordingly there is no directional change of the kind shown on paper by the line in Fig. 7.10. Consequently, we can state: The eye strides across the environment and its movement is such that (at least in the range of the axis of gaze) objective straight lines over which the gaze is passing are not interrupted by bends. Hence, this represents a further new type of active correction which has already been incorporated by the nervous sys-

tem in the gaze movement itself. Of course, this is also an innate mechanism, and not a learned effect.[1]

We have now arrived in the middle of the discussion surrounding a further constancy phenomenon − that of *form constancy*. But the functional principles involved in this phenomenon are still so obscure that we shall not consider it further here.

Hopefully, what has been said above has clearly demonstrated that the organism − as a perceiving subject − does not rely passively upon the physical influences in its environment. Only the sense organs themselves are passively exposed to the inflow of physical processes; but on the way to the perceptual level a process of *objectivation* operates. On the one hand, afferences from different sense organs are played off against one another, and on the other hand the particular stimulus configuration is evaluated for the purpose of objectivation. In many cases where a motor component is involved in the perceptual process, the central nervous process accompanying the relevant movement (or the movement itself) ensures the vitally necessary correction. It is consequently evident that perceived reality is quite different in composition from the reality investigated by the physicist, who records what reaches the sense organ from the exterior; briefly stated, the former is more *correct*![2]

This also applies − as is widely known − to the perception of colours. There is also the phenomenon of *colour constancy*; the ability to see colours 'correctly' independently of the accompanying lighting conditions at any given time. This ability is not particularly well developed but is adequate for survival. It is therefore tempting to try to summarize the various peculiarities of the organism, which operate in the service of this constancy function, in terms of a single functional system. Such an attempt (which of course suffers from

[1] The fact that a quite similar motor process − 'compensatory eye movement' − is involved (to a limited extent) in the end-result ensured in 'vertical constancy', was omitted in section on pp. 194–6 in order to simplify the description.

[2] It is perhaps worthwhile to underline the fact that this entire viewpoint should not be developed in the direction of 'naïve realism', according to which the external environment is constituted in exactly the same form as that which is perceived. This is not possible, because we are here confined within the realm of *what is perceivable*, and even the constancy functions themselves are only aimed at rendering the various possibilities of external experience open to interpretation with respect to an unchanging, objective world. This says nothing about the possible constitution of this world independently of the perceiving subject.

the fact that one cannot offer coloured illustrations here) will be made in the following account – with the accompanying danger that the reader will regard this as a mere display of intellectual acrobatics.

There are several phenomena in the field of colour perception in which our colour *vision* differs from the physics of light of different wavelengths. First of all, there is the fact that for us the colours are closed into a *circle*: this passes from red (which corresponds to the longest wavelength of visible light known to the physicist) through orange, yellow, yellow-green, green, green-blue, blue, blue-violet, violet and purple-red back to red. For the physicist, violet light has the shortest wavelength in the visible range; the colour purple does not exist in physics and arises in perception when light of extremely short wavelength is presented together with light of very long wavelength. For the physicist, light is arranged along a scale in which, as in the tonic scale, the longest and shortest wavelengths are separated by the greatest distance. A further peculiarity of our colour perception is the existence of a colourless, uniformly indifferent colour – *white* – which appears when light occurs in the same proportional spectral mixture as that fortuitously emitted by the sun. The colour circle, the colour purple-red and the colour white only appear to be as curious as they really are when we think of sound perception and try to imagine how the perceived tonal world would be constituted if the lowest sounds passed through transitional forms to the highest and if – in correspondence with the colour white – there were a pure sound without pitch.

In addition to this, there are further peculiarities in colour perception. The colours are grouped in pairs in such a way that when both members of a pair are presented with a particular intensity ratio they add up to give white, or 'extinguish' one another. These are the so-called *complementary colours*, such as a certain red with a certain green, or a certain blue with a given yellow, and so on. There is also the phenomenon of *colour contrast*, whereby the complementary colour becomes visible in the vicinity of a coloured spot, and the effect of *fading* of a colour when it fills the entire visual field or a major part of it. Finally, there is the correlated effect of the coloured *after-image*; that is, the fact that, after prolonged observation of a green picture, a subsequently observed white surface (for example) has a red appearance.

All of these peculiarities fall into *one* framework if one sets out from the concept that our colour perception apparatus has to cope with the task of solving the problem of *colour constancy*; in other words, it must ensure that we can recognize the colours of objects independently of the colour possessed by the light illuminating them and reflected by them.

It is evident that this task is not easy to solve; but living organisms have, after all, had many millions of years to do so. However, let us instead think of an engineer who is presented with the same task. This man must, therefore, carry out corrections on the light projected onto objects with the aim of cancelling or compensating any changes occurring with coloured illumination of some kind. To do this, it is first of all necessary to have a valid hypothesis for determining the particular colour of any illumination, since one cannot of course always go off in search of the light source. The most obvious working hypothesis runs as follows: the more a particular colour predominates in the visual field, the more likely it is that this is the colour of the illumination. It is, after all, improbable that many different objects will possess the same colour. It is therefore necessary to find a construction which will eliminate the predominant colour in the visual field to a degree proportional to its predominance. But, in the process, it is not permissible to interfere with the view of objects; that is, the undesirable colour cannot be replaced by nothing (darkness), since the undesirable distortion would be replaced by even greater disruption. Only the colour itself should disappear; the objects must remain visible. This task can apparently only be solved first of all by *inventing* a colour which possesses no colour value (white); secondly by randomly coupling every colour with another, complementary one which will extinguish it; and thirdly by providing the apparatus with the possibility of actively producing the complementary colour in order to eliminate the undesirable colour. The question as to which should be the so-called white, neutral colour and which colour should combine with which complementary colour to give white is open to arbitrary calibration. The simplest solution would be to establish as 'white' the colour mixture normally radiated by the light source; in other words, to give the property 'white' to sunlight, since this would ensure minimum utilization of the correction mechanism. The question as to which pairs of colours should be complementary could be solved in various ways.

One could pair the longest wavelength with the shortest, the next longest with the next shortest, and so on, but then there would be difficulties in the centre of the wavelength range, which would require complex auxiliary mechanisms for their resolution. Alternatively, one could combine light of the longest wavelength with a colour in the central wavelength range, the next longest with the colour immediately following the centre on the short wavelength side, and so on. In this case too, natural difficulties arise as soon as one comes to the end of the spectrum. These difficulties are in any event unavoidable – unless one closes the colour scale to form a *circle* so that colours evoked by short wavelength and long wavelength light are combined together. If this is successful, one has an arrangement in which an antipode can be determined for every point without 'blockage' at any point.

If the complementary colours are determined by a calibration, it is still necessary to prescribe the speed of the process of complementary colour production and its spatial limitation. The eradication of a colour dominating the visual field should not occur instantaneously, since this would exclude identification of the blueness of the sky or the greenness of a large meadow on which the gaze happens to fall momentarily. Since the red light of the evening only emerges slowly, and since, when a path is leading through a forest thicket, the illumination through the green leaf cover also changes slowly with each step, the process can be allowed to take place slowly. And since it is only slowly brought into operation, it will also decay slowly. This means a trailing effect of the self-manufactured complementary colour – an unavoidable, but also insignificant, disruptive effect (the coloured after-image) which must be accepted. In the spatial limitation of the effect, any sharp delimitation must also be avoided, since the mechanism is not concerned with neutralization of a localized colour spot, but with achievement of a general conversion of the colour values in the direction of the condition with white daylight. This general, diffuse response is, in any case, favoured by continual movements of the eyes, and its insignificant side effect, that of colour contrast, can equally be accepted, since it does not hinder clarity of the objects.[1]

[1] The widespread view that colour contrast increases the clarity of viewed objects is presumably erroneous. It is forgotten that the saturation of the colour of an object decreases to the same degree as that to which the complementary colour appears in the environment, since both effects are, of course, based on the same process.

The details of the mechanism which the technologist would design need not concern us here. (It is also admitted that this solution is not the only conceivable one.) Nature has in fact followed this route: the colour circle, purple and white, complementary colours, the after-image and colour contrast are properties of our perceptual apparatus. And the stimulus of this conceptual approach arises less from the fact that it is compelling than that it combines so many phenomena in *one* conceptual picture. The decisive function – *active manufacture* of a complementary colour – is evident to anyone who enters a room illuminated by a red lamp, where everything initially appears to be red-tinted and then after a while the individual objects once again assume roughly the same appearance as that seen in daylight. If the red lamp is now replaced by a white one, everything is immediately seen to radiate the green colour with which *we* have (subconsciously) painted the visual environment in order to extinguish the effect of the red light source.

The manner of setting questions and explaining uses in this article is not the same as that commonly employed in physiology and psychology. When we succeed in making phenomena plausible in one systemic context – and, in individual cases, in providing the system with a specific form, so that verifiable predictions are possible – this nevertheless leaves a wealth of individual investigation of the operative factors to be carried out. Such detailed causal analysis cannot, indeed, replace our systemic concept; but it provides the meat which fills out the conceptual carcase with which we have been exclusively concerned here.

Nevertheless, it should be said in conclusion that it was not my primary concern to demonstrate this conceptualization in terms of dynamic systems. My actual goal lay elsewhere.

The science of behaviour of living organisms has still not freed itself from the old chains of presumed (physical) exactitude imposed by the dogma that the activity of organisms is entirely determined by external stimuli – that it operates 'reflexly'. Any assumption of inherent active forces is dismissed as disguised vitalism and is therefore forbidden to any serious investigator. In view of such superstition, I regard it as particularly rewarding to demonstrate that *active, spontaneous processes* are not only present in active behaviour, but even in *apparently passive* scanning of the environment. *Without these processes, the organism would not even be able to* 'respond' *in an appropriate manner.* These processes are, in fact, an age-old hereditary possession. With animals, we are forced to conclude that they are present from an analysis of their behaviour; with ourselves, we can directly observe them, at least in some instances. Perception itself convinces us of the fact that internal processes are continuously operating within us to move what is seen to and fro, to make it larger or smaller, to alter its proportions, bend its lines and to paint the world with bright colours. It is thanks to such forces that we do not respond to 'stimuli', but instead see *objects* and are able to recognize them in different situations.

8 On the functional organization of drives[1]   1960

WITH URSULA VON SAINT PAUL
IN MEMORY OF OUR GOOD FRIEND GUSTAV KRAMER
(KILLED 19 APRIL 1959)

Introduction: the problem of localization

The human mind always wants to order its environment. Each thing which differs from others in its application, mode of functioning and appearance receives its own place and name. This way of thinking has also proved itself in the study of vital processes, which in fact do take place in organs, differing from each other in performance, mode of operation, appearance and position. Only in the case of the central nervous system (CNS) does this anthropomorphic will to order fail. The achievements of this entity are exceedingly varied. To name only a few: retaining a particular temperature value as reference point for temperature regulation; reflex movements as a protection against damage from outside agencies; preservation of an extract of past sensory data for use later on; carrying out various seeking activities, corresponding to changing bodily requirements: adaptation of the visual process to different light intensities so as to ensure good perception of objects; 'filtering out' particular stimulus properties of the outer world, to which appropriate instinctive behaviour patterns can then be linked up. If the interior of a part of the body with so manifold capacities were only now to become accessible to investigation for the first time, we should expect to find there a large number of different organs. As is well known, in the CNS – unfortunately – the opposite is the case; we find no recogniz-

[1] This translation by J. E. Burchard, Jnr., first appeared in the *Journal of Animal Behaviour*, 1963, XI, 1–20.

able organ boundaries, but everywhere almost identical structural elements, the neurones, in staggering numbers.

This morphological state of affairs has continually tempted investigators to search for 'the' functional principle of the CNS. Thanks to its manifold capacities, moreover, this organ system is able to answer the most diverse experimental 'leading questions' with a conciliatory 'yes' – even to postulates which are mutually exclusive. In this way have arisen the extreme forms of the reflex doctrine, of the doctrine of the 'conditioned' reflex, the thesis that all actions are learned, the doctrine of centres and the doctrine of plasticity, as monistic exaggerations of partial truths.

For many investigators, now as in earlier times, the most important question concerns the place at which a function occurs, the *localization problem*. The *doctrine of centres* finds abundant evidence for an anthropomorphic order: the elimination of certain functions by local brain damage is proof of organization into centres with afferent and efferent pathways. The *doctrine of plasticity* provides abundant examples of just the opposite: the impossibility of finding a specific area, for instance for memory or for the coordination of movement. The doctrine of centres argues further: if local electrical stimulation elicits particular reactions, then centres can be 'functionally' defined in this way. The opponents of this view reply that afferent or intracentral pathways would then, grotesquely enough, become 'centres' at whatever point the electrode happened to strike them; for it is just such pathways which most easily initiate responses. Only slowly has the contest between these arguments led to abandonment of the anthropomorphic scheme of organization. One may think that it might be economical to distribute the neurones serving a given function over a wide area, and to intersperse them with those serving other functions – namely, when cross-relationships among these functions are called for.

The most rigorous attempt to clarify the localization problem for a particular region of the brain is due to W. R. Hess (1954), who, by stimulating with implanted electrodes and marking the stimulated points by electrocoagulation, created in the course of many years a histological atlas of the midbrain of the cat, with many stimulus points for various reactions. The result brings no decision in favour of either of the two doctrines: many reactions could be elicited over a wide area, though always from specific places, between which sites for other reactions occurred; yet there were also brain regions for only a few functionally interrelated actions.

Programme and methods

Our own experiences with the internal dynamics of the CNS (von Holst, 1939c, von Holst and Jeschorek, 1956) and tentative experiments with a stimulation method derived from that of Hess have led us onto another methodological path. They confirmed a long-standing suspicion that the investigation of the physiological questions 'How?' and 'Why?' is usually deflected much too soon into the morphological question 'Where?' (von Holst, 1935g). Thus the histologist is burdened with problems which only the physiologist can solve by delving deeper into the central dynamics. We have therefore formed two distinct research teams, one of which studies only the central functional organization, the *Wirkungsgefüge*,[1] while the other studies the histological structure.[2] Not until the physiological method is confronted with insurmountable obstacles do we plan to attempt to correlate the functional with the histological organization.

Our experimental object is the domestic fowl, with its wealth of gestures and calls, its drastic sexual dimorphism, and its social structure. As to method, let it suffice to say that insulated silver wires, 0·2 mm in diameter and with uninsulated tips 0·3 to 0·5 mm long, are pushed into the brain in calibrated steps by means of a small screw arrangement made fast to the skull. Of the four (sometimes eight) electrodes, some are always left as indifferent electrodes in the 'silent' roof of the brain, while the remainder are used to seek effective fields of stimulation in the depths (Fig. 8.1.); it is impossible to say how large or of what form these fields may be, since the CNS is a *non-homogeneous* conducting mass. It cannot be taken for granted, furthermore, that the structures excited always lie in the immediate proximity of the electrode tip; when, for example, the tissues surrounding the electrode are destroyed by electrocoagulation

[1] The expression *'Wirkungsgefüge'* (control pattern or functional organization – Gefüge = structure, system; Wirkung = effect, operation) was proposed by H. Mittelstaedt (1954); we had previously spoken of 'functional structure' (*Funktionsstruktur*) (1939b).

[2] Under the leadership of W. Hirschberger.

for a distance of 0·3 to 0·5 mm, the previous reaction may still be present when stimulation is renewed.

The intact and unrestrained animals – which can be used again

Figure 8.1. Radiograph of a hen's head from the side, in which are implanted two brain electrodes. The small dark square above the ends of the wires and the triangle above right mark the auditory passages; the upper quadrangle between the four electrodes marks the end of the (invisible) Plexiglass electrode holder, which is screwed to the roof of the skull. Important areas of stimulation can be roughly localized with the aid of such pictures taken in two directions.

and again for years without the slightest ill effect – are free to move about on a table.[1] From the brain region indicated in Fig. 8.2 nearly all known movements can be elicited, serving orientation and the needs of the body, directed towards enemies, rivals, the sex partner and the young, with all the associated calls.[2]

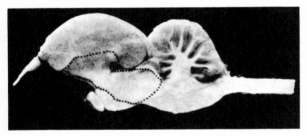

Figure 8.2. Longitudinal section of the brain of a fowl. The areas of stimulation explored lie in the region of the brain-stem marked with a dotted line.

[1] A method of wireless stimulation has also been developed (von Holst and Jeschorek, 1956) but will not be discussed here.
[2] A film showing the method used, and several films showing move of the reactions obtained as well as some experiments discussed later on, have been published by the Institut für den Wissenschaftlichen Film in Göttingen.

Grounds for confidence and lack of confidence in localization

Fig. 8.3 shows examples of how, when stimulating with sinusoidal alternating current, the threshold voltage at which the reaction begins depends on the frequency. The threshold has a minimum at 50 to 100 cycles per second, and climbs steeply with decreasing frequency, less steeply with rising frequency. One could thus, using a constant high voltage, ascend from 0 cycles or descend from 2000 cycles, and measure the threshold in cycles; or on the other hand – as we have done – express the threshold in volts at a constant frequency of 50 cycles.

Let us trace, with such threshold measurements, the path of an

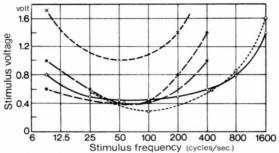

Figure 8.3. Examples of the dependence of the necessary stimulus voltage (threshold voltage) on the stimulus frequency. Sinusoidal alternating current. Reactions: x and △ = clucking; ● = looking about (*Aufmerken*) with jerky head movements in all directions: o = calling to food; ■ = watching out (*Sichern*) with extended neck and motionless head.

electrode as it penetrates in tiny steps downwards somewhere in the brain stem. One should expect that as it passes by a localized structure the threshold value would be high at first, then low at the point of closest approach, and then high again. This is, in fact, often the case; Figs 8.4 and 8.5 give examples. In many other cases the threshold curve is irregular and extends over as much as several millimetres (Fig. 8.4 bottom, ● □); which perhaps means that the electrode is travelling parallel to a fibre tract. When the stimulus voltage is increased, other reactions usually appear in addition to (or instead of) the first (Fig. 8.4, bottom); this too is understandable, since the region of adequate stimulation has expanded to affect new structures.

224

Figure 8.4. Examples of the appearance and disappearance of reactions as the electrode is pushed forward in the brain stem. The horizontal markings along the electrode path designate millimetre intervals. The horizontal distance of the points from the electrode indicates the threshold voltage (see scale below). The reactions are: ● = looking about (*Aufmerken*); x = watching out (*Sichern*); + = head-shaking; o = turning to the left; □ = turning to the right.

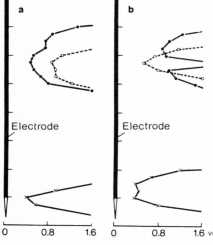

Figure 8.5. Another example of the appearance of different behaviour patterns as the electrode moves deeper. Details as in Fig. 8.4. ● = looking about (*Aufmerken*); o = turning to the left; x = fleeing downwards. In *a* the electrode was moved downwards; in *b*, about 30 minutes later, it was withdrawn upwards along the same path.

Less illuminating is the common situation that the same reaction can be obtained from various fields of stimulation; in the intervening area, other behaviour patterns are elicited (Fig. 8.4, ●). This fact, well known in the literature, is difficult for the histologist to interpret; we shall see, however, that it has a physiological explanation. Completely mystifying for the histologist, finally, is the common phenomenon that the electrode, moving for a second time along the same path, produces different behaviour, even with the same stimulus strength. A relatively innocuous example of this is shown in Fig. 8.6, where at one location a turning movement appears in place of looking around (*Aufmerken*). If one leaves the electrode in one spot and extends the experiment over several hours, it is not unusual to obtain at times nothing, then perhaps running away a couple of times, later on clucking, after that fluffing out of the feathers, and finally perhaps preening – all from the same field with the same stimulus voltage. This blow to the proponents of localization, too, follows from rules of the functional organization, which we shall come to know.

Precision of the method; the compensation formula; measurement of visible and invisible phenomena

The behaviour of a higher organism is the result of a complicated interaction of qualitative and quantitative data; is the method at all accurate enough for the task of measuring these data? The facts just mentioned encourage scepticism.

Many things might be quantified: the necessary threshold voltage, the latency with which the reaction follows the beginning of stimulation, the intensity of the reaction (i.e. speed or amplitude of a movement, volume of a sound), the frequency in cases of rhythmic response, and finally the duration of the reaction – always with the condition that the *quality* of the behaviour remains *constant* throughout the series of measurements.

Let us look at some examples to confirm the regularity of the phenomena. For movements with a well-defined beginning, a good quantity to measure is the *latency*, whose relationship to the *stimulus voltage* is shown in Fig. 8.6 for the example of the aerial-warning cry. Here, as in other cases, the values lie – other things being equal – on a hyperbola-like curve; as the voltage rises, the necessary stimulus duration sinks. The product of these two quantities would be constant in a hyperbola; here, however, it is constant over a wide range only if we first subtract a certain amount – about two-thirds

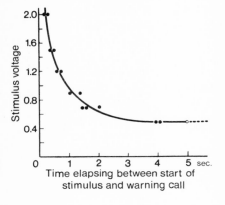

Figure 8.6. Dependence of the latency on the stimulus voltage for the aerial-warning call ('*Klock*') of a cock. Here and in all further measurements, sinusoidal alternating current of 50 cycles/sec. has been used. (The stimulus voltages were varied at random in the experimental sequence; the first stimulus which released no reaction is indicated by o.)

of the lowest threshold voltage, the 'rheobase' – as shown for three such experiments in Fig. 8.7. Only with very high stimulus voltages, at the left-hand end of the curve, is there a downward bend; here the product becomes smaller, which is to say the stimulation is more effective.

It seems as if we might here be dealing with a rather general law; this cannot be tested accurately with the latency method, how-

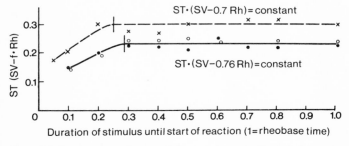

Figure 8.7. For three experiments of the type shown in Fig. 8.6, the product of the stimulus time (ST) times the corresponding stimulus (SV) minus a part of the lowest effective stimulus voltage, the rheobase, (f. Rh.), is plotted as a function of the stimulus duration. (x = aerial-warning call of the cock; ● = headshaking; o = 'alert call' (*Wachlaut*) – a brief, explosive burst of cackling.)

ever, in cases where the behaviour begins gradually, for instance sitting down or turning the head to one side. Here a methodological trick is useful: the behaviour patterns activated in the brain stem continue smoothly if the stimulating current is rhythmically interrupted (between 4 and 10 cycles/sec.). It is then possible, using a constant interruption frequency, to vary the length of the stimulus pauses and measure the corresponding rheobases, that is the minimum voltages necessary to produce any reaction at all. These values are correlated with the ratio of stimulus length to pause length and are independent of the latency; the latter disappears from the equation. Fig. 8.8 shows nine series of experiments conducted in this way; all the curves are quite similar. When they are reduced to the same initial value, averaged, and plotted similarly to Fig. 8.7, the curve shown in Fig. 8.9 results; it is in full agreement with the results of the latency measurement as shown in Fig. 8.7. We have thus found a general formula describing the way in which the intensity and the duration of the stimulation substitute for one another, or compensate each other: the *compensation formula*.

The measurements thus far have used thresholds of behaviour; these thresholds vanish when the stimulus-dependent activity is already present spontaneously, as Fig. 8.10 shows for the jerky head movements in looking around (*Aufmerken*). Here the frequency gives a good measure with (other things being equal) reproducible data; one can see that the curve tends toward a 'saturation' value of about three movements per second.

Fig. 8.11 shows, finally, with sitting down as example, the dependence on the stimulus voltage of four different quantities: latency, speed, amplitude and duration of the movement. We shall consider for the moment only the striking phenomenon of *after-response*, here dependent on the stimulus strength (and thus, according to the compensation formula, also on the stimulus duration) (Fig. 8.11b). What is the immediate cause of this rather common continuation of the activity long after the end of stimulation? Has the stimulus piled up some specific something centrally, which is slowly 'used up', and can this invisible something be measured?

The student of behaviour knows that actions which are externally the same can be internally of very different intensity. A sitting hen can be slightly or strongly 'inclined to sit'; in the first case she can be distracted easily by other stimuli, in the second case with difficulty.

228

The same is true of many activities, the degree of whose necessity or 'drive' is not outwardly apparent, such as standing, drinking, sleeping, preening, picking off of vermin, crowing, etc.[1] One can,

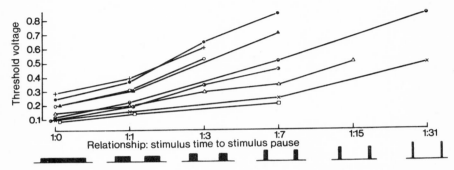

Figure 8.8. Measure of the threshold voltage for continuous stimulation (alternating current, 50 cy/sec.) and for rhythmically interrupted stimulation with increasing length of pauses and decreasing length of stimuli, as the sketch below suggests. (x = headshaking; △ = turning head away; o = sitting down; the other six symbols indicate clucking.) The frequency of stimulus interruption, between 4 and 10 cy/sec., is without effect.

however, always measure this 'drive' by means of a second stimulator and a second electrode, with which one seeks out the stimulus field of an opposing response. To make clear the principle, the simple scheme of Fig. 8.12 suffices, in which sitting down and stand-

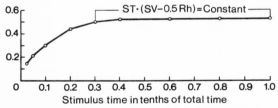

Figure 8.9. The values of Fig. 8.8 are summarized, recalculated as in Fig. 8.7 according to the formula: ST.(SV–f.Rh), and plotted as a function of the ratio of stimulus time to total time (thus 1 is continuous stimulation, 0·5 indicates that stimulus time equals pause time, and so on).

[1] In order to make more data visible, we have sometimes also recorded respiration and heartbeat; nothing further will be said here of these, nor of the changes in action potentials which can be recorded from the electrodes during the pauses in stimulation.

ing up were played off against one another: at the same time this scheme brings all the quantities so far discussed into a natural relationship.

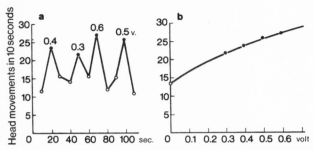

Figure 8.10. Spontaneous jerky head movements in all directions – looking about (*Aufmerken*) – are increased by stimulation. *a* shows the frequency increase of the head jerks in the sequence of the experiment. *b* the same in systematic arrangement. The figure does not show that during stimulation the head jerks are directed forward and merge into intention movements of picking at little particles on the ground or on the bird's own body. (o = spontaneous frequency; ● = frequency during stimulation.)

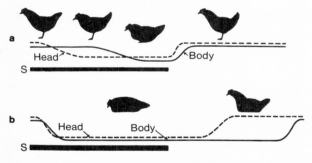

Figure 8.11. Effect of two sitting stimuli of different strengths on a quietly standing fowl; pulling in the head and lowering the body are registered separately; the stimulus (S) amounts to 0·5 volt in *a*, 0·7 volt in *b*. (From cine film.)

The subject is a hen with spontaneous 'sitting drive', which we cause to stand up by stimulation of a central 'standing field'. The first stimulus (Fig. 8.12a) is strong but short; the animal stands up quickly after a brief latency, and sits down again at once. The second stimulus (Fig. 8.12b) is strong and longer; latency and standing up are unchanged, but the standing persists for a time after the end of stimulation. The third stimulus (Fig. 8.12c) is weaker and much longer; the latency, following the compensation formula, is longer, the hen stands up slowly, and the standing persists, again following the compensation formula, just as long after the end of stimulation as in the second case. As common cause of these experimental facts a central process may be postulated, indicated by the curve running across the middle of the figure; it has the value zero at the point where sitting changes to standing; below the zero line we call it physiological *sitting drive*, above the line *standing drive*. This postulated central process is, in fact, directly measurable: as long as the

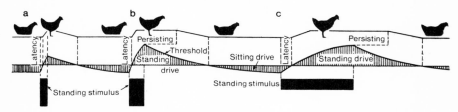

Figure 8.12. Sketch to illustrate the interrelation of various parameters of the behaviour. Explanation in text.

animal is sitting, the threshold voltage for standing up gives a measure of the sitting drive that must be overcome; as long as it is standing, on the other hand, the threshold voltage which must be applied at any particular moment to a central 'sitting field', in order to induce sitting down, is a measure of the standing drive. The schematic course of the whole threshold curve is drawn according to such measurements using two complementary stimulation fields. It defines, first of all, the magnitude and direction of the central drive – whose physical nature is of course completely obscure; secondly, it determines in a readily apparent way the visible data: latency, speed, duration and their relation to stimulus strength and duration. And in so doing, thirdly, it explains the principle of the compensation formula.

An excursion on the precision of the method

The previous section has made clear that the method can satisfy high requirements as to exactness and reproducibility. But under what conditions? What, for example, does the frequently cited qualification 'other things being equal' mean? The answer is this: the method is exact only so long as the *internal* central situation remains constant, in which every action is, as it were, embedded. Many and various factors, temporally often far removed, influence this situation; we shall become familiar with the most important.

How does one attain a constant – or at least known – *internal* situation? It is no use grabbing an animal from the henhouse, inserting electrodes (under anaesthetic), and then, when it wakes up, beginning the stimulation experiment: most of the stimulation fields remain silent, while from others one gets mainly 'freezing' and various sorts of fleeing: reproducible curves are unthinkable. If one succeeds, in the course of hours, in getting the animal to settle down somewhat, many of the silent regions will become active, and fleeing will become confined to fewer stimulation fields; if one is fortunate enough to 'put over' perhaps a brooding, sleeping, or crowing reaction, the ice is generally broken for the time being. Precondition for all measurements is a calm, 'comfortable' basic mood, in the fowl very roughly characterized by loosely lying plumage, alert looking around, tendency to preen, to eat, to sit down, and in the cock also by spontaneous crowing, by watching out (*Sichern*) and (perhaps) warning and by calling to food. Even the slightest tension (tonic immobility or 'freezing') can ruin everything. To be able to differentiate such fine nuances in posture and movement, a knowledge of the full inventory of behaviour is just as essential *before* beginning the experiments, as it is afterwards necessary for the correct identification of every reaction.[1] Once the desired neutral basic mood has been approximated, one obtains, in the course of the next few hours, as more stimulus fields are reached, correspondingly more and more control of the whole situation, and one is then in a position to measure factors which *change* this internal situation; we will now direct our attention to these.

Central adaptation and mood shifts ('Umstimmung')

To the observer of animals it is a familiar story that a stimulus situation sets a particular action going with at first rapidly rising and

---

[1] We wish to thank our co-worker, the outstanding fowl specialist Dr Erich Baeumer, for introducing us to the very extensive catalogue of fowl behaviour; in all doubtful cases, we have relied on his cautious interpretation (Baeumer, 1955).

then gradually declining intensity, as if at first an 'initial friction' had to be overcome, and as if the behaviour later became 'fatigued'. This phenomenon also appears quite generally when a field in the

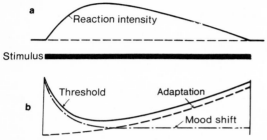

Figure 8.13. a: curve illustrating the rise and fall of a reaction with constant stimulation. *b*: initial decline and subsequent recovery of the threshold, for the same reaction as in *a*. The way in which this threshold curve results from two processes, the one increasing and the other decreasing, is explained in the text.

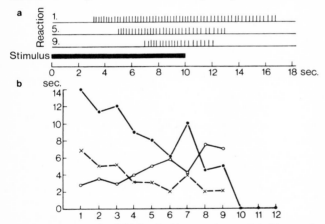

Figure 8.14. A stimulus field for clucking was 'pumped dry' by a series of twelve stimulations, each of 10 seconds duration (with intervals of 10 seconds between stimulations). *a* indicates the result of the first, fifth and ninth stimulations; *b* gives for the twelve stimuli the latencies (o), the persistence (x) or after-response duration, and the duration of the response (●).

brain is stimulated (Fig. 8.13a). If one measures it in terms of stimulus thresholds, the threshold starts out high, then falls rapidly, and later climbs slowly up again (Fig. 8.13b).

We will first examine the 'fatigue' phenomenon. Fig. 8.14 shows the fading away to nothing of a clucking reaction in twelve successive identical stimulation series; one can see here how the shortening of the response duration comes about through rising latency and symmetrically corresponding decline in after-response. In the experiment shown in Fig. 8.15, the same clucking is a component of a more complex fleeing behaviour. Here, following the scheme of Fig. 8.13b, the stimulus was increased parallel to the slowly rising threshold; thus not the stimulus strength, but the reaction was held constant. Here it is abruptly ended by the appearance of a new behaviour element, flying away.

All such cases of decrease in reaction (or of increase in threshold) have one thing in common: they give the same picture as would

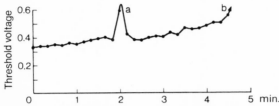

Figure 8.15. A fowl is brought into fleeing restlessness through stimulation (the behaviour elements which appear are shown in Fig. 8.29a). The continuous stimulation was gradually increased in such a way that the animal was constantly clucking softly; the stimulus voltage was read off every 10 seconds. At *a* the bird defecated; at *b* it flew off with a brief screech.

be given – other things being equal – by a decrease in *stimulus voltage*. One might suppose that the conduction path from the stimulus field to the neuromotor apparatus proper is being increasingly blocked at some point or other.

In favourable cases it can, in fact, be shown that this suspicion is correct. For a simple and common reaction, for example clucking, we search out two different, widely separated stimulation fields. With two independent stimulators we can now carry out a variety of stimulus combinations (Fig. 8.16), from which many kinds of information can be obtained. We can, for example, apply to each field a voltage, which produces a moderate reaction; then both fields are stimulated simultaneously, and there appears, as expected, an increased reaction (Fig. 8.16b). There is thus no doubt that the two excitations flow together and are summated *somewhere* in the CNS. Now we stimulate one field alone for a longer time, until after the behaviour has faded out, and then at once stimulate the other field alone; the reaction immediately reappears in full force, and vice versa (Fig. 8.16c). First of all, this confirms that we are dealing with two distinct physically non-overlapping fields of stimulation; and it further shows that the fading out takes place *before* the two excitations flow together, since otherwise the effect would necessarily still be noticeable in the *other* stimulation field. It follows, then, that less and less excitation does in fact get through to the neuromotor appar-

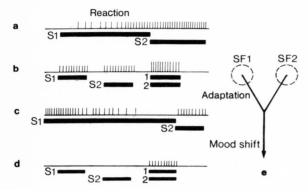

Figure 8.16. Scheme to explain various combinations in the stimulation of two fields for the same behaviour. Based on a case in which the two stimulation fields lay in the right and left brainstem halves, laterally separated by a distance of about 5 mm, and gave 'pure' clucking. To be certain of excluding a possible mutual physical interaction due to voltages produced in the tissue by the two pairs of electrodes, when both fields were stimulated simultaneously, rhythmically interrupted stimulation was used (as in Fig. 8.8) in such a way that when both fields were stimulated the stimulation in one occurred during the pauses in the other. (S = stimulus; SF = stimulus field; further explanation in text.)

atus. We therefore designate the phenomenon a central, local *adaptation*.

If this is all true, it has still another consequence: when a field is continuously stimulated, this afferent structure should *remain* continuously adapted, so that the extinguished behaviour does not reappear no matter how long stimulation is continued. This is, in fact, regularly the case, and expresses itself particularly drastically for reactions which adapt very quickly so that stimulation results in but a single 'discharge', for instance crowing, the aerial-warning cry, the 'alert call' or 'Wachlaut' (a brief 'burst' of cackling). Thus, it is not unusual for a cock to crow but once with continued stimulation of a particular 'crowing field', but if the stimulation is briefly inter-

235

rupted after each call, it may crow 20 times or more within 5 minutes! This shows very nicely that a 'fatigue' of the neuromotor apparatus itself is out of the question, since – with much less total amount of stimulation – it is so much more active in the second case. By varying the length of the pauses between stimuli, it is possible to follow the process of *de*adaptation; in the experiment in Fig. 8.17 it required something more than 10 seconds.

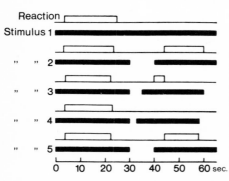

Figure 8.17. Dependence of a reaction – cleaning tongue movements in the mouth – on the stimulus duration and stimulus pause. The stimulus voltage is constant (0·5 volt). With prolonged stimulation the reaction stops after about 20 seconds, and the occurrence and duration of its reappearance depend on the length of the pause in stimulation.

These data once again show that it is possible to measure something that is not directly visible; here it is not a drive, but a localized process in a conducting structure. The existence of such *subthreshold* events is, as we shall see, important for the localization problem; it can be demonstrated still more directly in the experiment with two stimulus fields for the same reaction. In many cases, namely, when both fields are stimulated simultaneously by voltages each of which alone is wholly ineffective, a clear reaction occurs (Fig. 8.16d). This means, however, that both 'subthreshold' excitation processes must be transmitted at least as far as the place where they come together, since otherwise they could not summate and so raise the behaviour itself over the threshold. Thus the threshold, which must here be overcome, does not lie in the stimulation field at all, but *somewhere* on the way to the motor apparatus.

Now let us consider the other phenomenon: the decline of the threshold at the beginning of a long-continuing or frequently repeated action (cf. Fig. 8.13b). As Fig. 8.18a shows, this effect can be absent entirely when the stimulus is infrequent (or short); the threshold curve remains on one level. The more closely the stimuli follow one another, the more steeply does the threshold curve fall

– here even to zero (Fig. 8.18c); that is, the *re*action becomes a *spontaneous* action. In other cases, the latency can serve as measure instead of the threshold (according to the compensation formula); here too, tremendous initial changes can appear, for which Fig. 8.19 gives an example.

In surveying numerous similar cases, it will become clear to the observer that the curves only show *one* side of the situation, for every

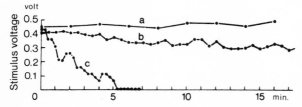

Figure 8.18. A fowl sits motionless (sleepy) in dim light. A field is stimulated which produces looking about (*Aufmerken*), with a stimulus which increases (by 0·1 volt every 3 seconds) until the first head movement; then the stimulus stops instantly. In *a*, stimulation was every 90 seconds; in *b*, every 30 seconds; in *c*, every 15 seconds.

such decline in threshold is accompanied by a simultaneous rise in the thresholds for certain other activities. We are dealing with the displacement of an equilibrium condition, along the lines indicated in Fig. 8.12, except that here much slower processes are involved, which are overlain by the threshold relationships of the activity itself, without necessarily being influenced by them. Thus, for instance, in Fig. 8.20 a long-sustained scolding mood is interrupted from two stimulation fields. One of the stimuli – watchful staring (*Sichern*) – has only a slight influence, the other – sitting down in sleepy mood – is more effective; with brief stimulation, nevertheless, the scolding mood is not yet altered. After a prolonged sitting stimulus, however (Fig. 8.20b), the hen stands up and is quite pacified. One sees, as in Fig. 8.18a, that the basic mood has an inertia and requires time to change. We designate such gradual changes of a basic tendency as *mood shifts*.

In Fig. 8.18b, c and Fig. 8.19, looking around (*Aufmerken*) and clucking are signs of a mood shift from sleep to a wakeful condition. A sleep stimulus would now be less effective. With other stimulus fields, one can similarly shift from hunger to thirst or fleeing or brooding and vice versa, or from courting to fighting mood, etc., but it is unfortunately not possible to arrange the various moods without violence into antagonistic pairs. Like many threads leading from a single knot, they pull in various directions; a fully neutral

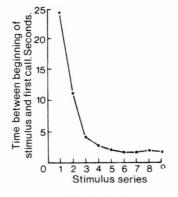

Figure 8.19. A quietly sitting fowl was repeatedly stimulated with a constant stimulus (0·4 volt) in a field for clucking, each time until the first cluck. The stimulus duration (latency) was measured. Between one stimulus and the next was a pause of 5 seconds (only the first two stimuli, not shown here, had other voltages; 0·38 volt: no reaction; 0·42 volt: reaction after 3 seconds).

basic mood 'nil', furthermore, apparently does not exist.

With a single mood are associated several behaviour elements, which thus reveal an internal relationship. Thus spontaneous crowing, a general expression of masculine self-confidence, very often follows stimulus-induced 'wing-scratching' (*Kratzfuss*) (Fig. 8.21), or

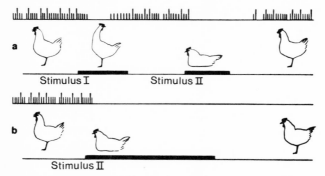

Figure 8.20. A fowl, which in consequence of a preceding fleeing stimulus cackles spontaneously and without stopping (kut, kut, kudaw, kut, kut, kut, kudaw! . . ., suggested by the upper graph). Stimulus I produces watching out (*Sichern*); Stimulus II produces sitting down; further explanation in text.

calling a hen to the nest, or threatening a rival, or pecking. The stimulus, so to speak, activates in each case a specific masculine activity, which ceases with the termination of the stimulus and leaves behind only the basic mood 'increased manliness', which finds expression in spontaneous crowing.

When a stimulus-induced mood comes to an end, it can also give way temporarily to its opposite. All the very varied reactions listed in Fig. 8.22, when they are activated for some little time and not too intensively, strongly promote crowing as a spontaneous follow-up. Looking through this list, one can designate 'moderate depression' as common element. This mood, then, when set free rebounds

Figure 8.21. Cock showing 'wing-scratching' as a consequence of brainstem stimulation (0·15 volts). This is a display in which the animal circles a subordinate companion, usually a hen, scratching with the outer foot at the downstretched outer wing. Here, in the absence of another fowl, it is a 'vacuum activity'.

into increased 'self-esteem'[1]. After very strong depression, however, crowing does not follow at once, but rather first cackling, as expression of the overcoming of strong tension.

All this gives the phenomenon of mood shifting a very distinctive character, quite different from central adaptation, and encourages the suspicion that it might occur at a different place in the CNS. Here again we have recourse to the experiment with two independent stimulation fields for the same reaction, and ask whether a mood shift produced from one field is also effective for the other or not, and vice versa. In the cases so far examined, as the scheme in Fig. 8.16a shows, the answer is: yes! This means that the mood shift, unlike the adaptation, takes place *after* the confluence of the two stimulus effects, and thus nearer to the neuromotor apparatus,

[1] According to the motto: 'Today I'll give my dog a treat; first I'll beat him – and then I'll stop!'

as Fig. 8.18c indicates. Thus, since two different processes – mood shifting and adaptation – always affect the behaviour threshold, there at once results the further important conclusion that a given

Watching out
Clucking
Evasion
'Alert call'
Crouching
Jumping
Aiming
Screeching
Cackling

Figure 8.22. List of stimulus-released behaviour patterns which are often (in many cases regularly) followed, 5 or 10 seconds later, by crowing. (The 'alert call' is a brief 'burst' of cackling and could be roughly translated as 'what the devil!' Aiming is a component of fleeing behaviour, cf. Fig. 8.29a.)

threshold value does *not* always mean the same central situation; for it is clear that much mood shift plus much adaptation can give the same threshold value as a little mood shift plus a little adaptation (see Fig. 8.13b).

Thus far we have considered adaptation and mood shift isolated from other phenomena. Obviously there are many superpositions. Thus Fig. 8.23 shows a case of adaptation combined with persistence of the reaction (after-response). We already know that after-response depends on the stimulus duration (see Fig. 8.12a, b). When, as here with watching out (*Sichern*), adaptation enters the picture, *no* after-response occurs either with no stimulation or with very long stimulation, that is with complete adaptation. Between these extremes, however, lies a maximum of stimulus effect, so that the values for after-response form a peculiarly bent-back curve.

Fig. 8.24a shows the effect of a mood shift on a series of experiments with changing latencies. The latency data form a family of curves, which with continued stimulation drift towards smaller and smaller values; here one can, by means of a suitable assumption about the form of the mood shift (Fig. 8.24c), easily remove this effect, so that the latency data come to lie on a common curve (Fig. 8.24b).

Closer familiarity with adaptation and mood shifting phenomena greatly eases the interpretation of changing behaviour sequences, which *apparently* have a somewhat arbitrary character. Fig. 8.25 reproduces such a case, where three identical protracted stimuli bring a cock from drowsiness into a tense-wakeful mood, which expresses itself differently each time. We interpret: the first stimulus

(Fig. 8.25a) produces to begin with a strong fearful tension, which fades with increasing adaptation, so that the clucking gradually changes over to cackling. The second time (Fig. 8.25b), the same stimulus has a lesser depressive effect and therefore rebounds, after its decay with adaptation, to the opposite mood of masculine self-esteem, and then (as we already know) to crowing. Finally, the third

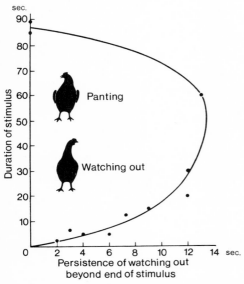

Figure 8.23. A fowl which is spontaneously panting (rapid breathing with open beak and wings held away from the body, performed when the body temperature is too high) is caused by stimulation to watch out (*Sichern*), whereby the panting posture rather suddenly vanishes. With a constant voltage (0·5 volt) the stimulus duration was varied; the persistence of watching out after the end of stimulation was measured (after-response). Explanation in text.

time (Fig. 8.25c), the stimulus effect is so weak that the quickly fading mood hardly rebounds to the other side at all; after a little clucking the cock drops back into his sleepy basic mood, while the stimulation is still going on.

Interaction of different behaviour patterns

Against the background of the slowly changing moods, the more rapid dynamics of the drives take place. One can study them experi-

241

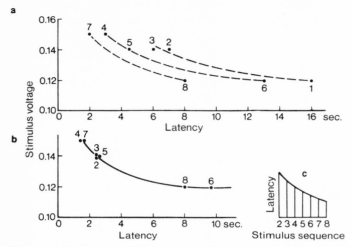

Figure 8.24. The way in which the latency of a reaction – jumping up and fleeing of a sitting cock – depends on the stimulus voltage *and* on the number of previous releases of the reaction. The numbers indicate the sequence of the stimulations (Stimulus 1 produced no reaction). In *b* it is assumed that the latency becomes shorter by one-sixth in each successive experiment (corresponding to the curve in *c*). Explanation in text.

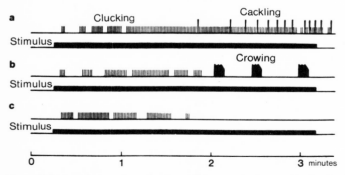

Figure 8.25. A cock, in a sleepy basic mood, is provoked to various kinds of activity (clucking, cackling, crowing) by three long, constant stimuli (0·2 volt) separated by brief pauses. Explanation in text.

mentally by simultaneous activation of two different behaviour patterns, either from two stimulus fields or from one field combined with an external stimulus. The various types of combination which can be distinguished are shown very schematically in the accompanying table.

Superposition is the simplest and least problematic form: both movements remain independent. For *averaging* this is no longer true;

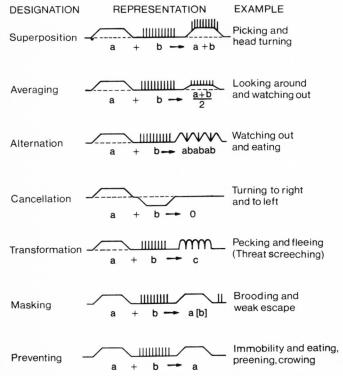

| DESIGNATION | REPRESENTATION | EXAMPLE |

Figure 8.26. Types of interaction of different behaviour patterns. For explanations see text.

thus combining rigid watching out (*Sichern*) with extended neck and attentive looking around (*Aufmerken*) with large head movements results in looking out with less extended neck and short, sharper head jerks. The more uncommon *alternation* appears when two tendencies are in equilibrium; a good example is eating and watching out, which together can produce an alternation of hasty food-pecking with brief, quick raising of the head.

The case of *cancellation* has already been explained with the scheme in Fig. 8.12 for standing and sitting. This type pertains only to exactly complementary behaviour patterns, such as turning to the left and to the right, which can cancel each other out to zero. For 'pure' sitting and 'pure' standing this case is actually realized, but for sitting as a part of brooding, standing as a part of food-seeking or of fleeing, the situation is certainly more complicated.

In all of these cases one cannot speak of a 'conflict'. A clear case of conflict, however, is the *'transformation'* which can appear with simultaneous attack and fleeing tendencies and which shows a new, strongly effective action: tripping about with puffed-out feathers and screeching. The behaviour is very reminiscent of the reaction of a hen on the nest, when something dangerous approaches her. The scheme of transformation, furthermore, well illustrates the draw-backs of forming types: it contains cases of diverse physiological makeup. For example, when one produces a slight fleeing tendency in a hungry animal, so that it ceases to eat, and then gives a sleeping stimulus, which counteracts the fleeing tendency, the animal may then eat for a short time before going to sleep. This would correspond formally to the scheme $a + b = c$ (fleeing + sleeping = eating); but it would be more correct to say: a suppresses c, b suppresses a; $a + b$ liberates c.

This brings us to the next type, *masking*, which is widely distributed. We use this term when one behaviour pattern makes a second one invisible, and it is nevertheless possible to show that the latent drive has not been eliminated. Thus, for example, sitting down (in broody mood) can completely mask a simultaneous slight clucking excitation. When sitting stimulus and clucking stimulus cease *simultaneously*, however, a few bursts of clucking still follow immediately. Or: fearful 'making-oneself-thin' masks feather-puffing-out; after simultaneous termination of both stimuli the plumage is puffed out for a few seconds. The masked action does, in fact, become visible as an 'after-explosion'; its drive must therefore have been present in latent form, but was blocked somewhere on its way to the motor apparatus. In most cases, unfortunately, where we suspect masking, such an after-reaction is not visible – which might indicate merely that the visible drive persists longer than the masked one. Here *'pre-*

venting' is the neutral term. Thus in the case of Fig. 8.20, where cackling is temporarily interrupted by watching out (*Sichern*) and sitting down, one can only speak of preventing; it would be clear masking if the hen cackled more vigorously afterwards. Fig. 8.15a shows another example of prevention; the defecation provoked by the fleeing stimulus momentarily greatly raises the threshold for clucking, likewise a consequence of the fleeing stimulus.

Which reaction is masked (prevented) by which other ones, usually depends on quantitative factors. We can, however, designate certain behaviour patterns as *dominating*, because they shut off any other activities even when only slightly activated. Such dominating

Figure 8.27. Dependence of the stimulus threshold for one action – picking to the left on the ground and on the body (I) – on the stimulus voltage with which a second action – standing motionless with head stretched forward, saliva flow, and long-persisting unwillingness to eat (II), the 'disgust posture' (compare Fig. 8.28, second sketch) – is simultaneously activated. Stimulus II (disgust posture) is first set at a particular value, then stimulus I is added and increased until the corresponding actions become visible. x = picking movements; ● = turning head to the left.

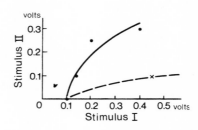

reactions are especially the various forms of fleeing, and still more 'freezing'[1] (*Starre*), against which other behaviour breaks through only with difficulty.

Measuring the possible *quantitative* effects of all behaviour patterns upon each other, through simultaneous activation in pairs, is an important step on the way to reconstructing the functional organization (*Wirkungsgefüge*). Fig. 8.27 gives an example, which shows several things: the one reaction is in this case a posture ('disgust posture' – *Ekelhaltung*) which does not change with increasing stimulation; pecking and head-turning have thus to contest against a *constant* posture of *varying drive strength*. One can see that increased 'disgust drive' raises the threshold moderately for head-turning, but steeply for pecking. Thus the 'disgust drive' seems more strongly contradictory to the pecking drive (which is closely related to the eating

[1] Freezing, so-called 'playing dead', appears in many birds in moments of deadly peril.

245

drive) than to the neutral sideways turning of the head, which can more easily be combined with the disgust posture.

On the whole such experiments indicate that among the individual behaviour patterns – as with the moods – it is generally not possible to set up antagonistic pairs; mutual masking (prevention) occurs in very varied gradations.

Finally, it may also happen that a stimulus does not raise the threshold for another behaviour pattern, but lowers it instead. An example: anxious excited clucking ('kut, kut, kut, kut . . .') of a hen in the farmyard is contagious among the other hens, but does not spread to the cock; he reacts to it with cackling ('kut, kudaw-kut, kut, kudaw'). Likewise, a tape recording of hens clucking produces at most brief cackling in the cock. If one activates in the cock a stimulus field for clucking, it can be seen that his clucking threshold does in fact sink, when he simultaneously hears recorded clucking; in fact, the louder the tape recording, the lower the stimulation threshold. We can therefore designate clucking and cackling as related (allied) reactions; as a matter of fact they do frequently merge into one another, for example in the case of Fig. 8.25a.

Complex behaviour sequences: the postulate of level-adequate terminology

Thus far we have concerned ourselves mostly with simple behaviour patterns, and will now take a closer look at the behaviour sequences into which they enter.

It has already been noted (Fig. 8.4, bottom) that one can very often obtain different behaviour from a single stimulus field with increasing voltage. From the manifold and varying movement sequences and mixtures of this sort, there clearly appears a type characterized by four features: (1) the sequence of the individual movements is unalterable; (2) one also obtains the sequence (generally) with a constant medium voltage; (3) the sequence is identified as *natural* by the behaviour specialist; (4) the sequence as a whole serves *one* particular function. We conclude from this that we are here stimulating one and the same structure, which activates a drive whose increase excites the individual behaviour components sequentially, corresponding to their individual thresholds.[1] We shall discuss a few cases.

One particular stimulus field produces in a cock blinking of the left eye at first; if the stimulus is prolonged or increased in intensity, occasional headshaking appears, then wiping the head against the shoulder, and finally scratching the left cheek with the foot. Head-shaking and scratching are repeated several times if the stimulation is prolonged enough. It looks as if the cock were being bothered by an invisible fly. Fig. 8.28 shows another case. The whole sequence

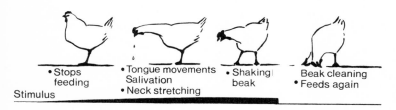

• Stops feeding

Stimulus

• Tongue movements
Salivation
• Neck stretching

• Shaking beak

Beak cleaning
• Feeds again

Figure 8.28. Activation of a behaviour sequence from one stimulus field. The whole complex serves to remove something unpleasant from the beak. The individual actions, except beak cleaning and salivation, can also be obtained in isolation from other fields of stimulation.

here imitates the disgust reaction on mistakenly picking up a (stink-) bug; beak-wiping *after the end* of stimulation is the normal and meaningful conclusion of the cleaning.

In both cases one presumes that a central pathway ascending from the cheek or mouth cavity lies in the stimulation field. Fig. 8.29a and Fig. 8.30a give examples of fleeing behaviour from a (non-existent) ground or aerial enemy. Here the stimulus field will lie in the neighbourhood of 'higher' pathways, whose activation effects the physiological correlate not of a sensation, but of a form of hallucination. Other cases correspond to the activation of specific bodily requirements: Fig. 8.31 shows the behaviour sequence of falling asleep, or the effect of the 'sleep drive'. In the same category belongs the activation of searching for water with subsequent drinking, or search for food with eating, or for instance (in the cock) searching for a

[1] A view long maintained by outstanding students of animal behaviour (Whitman, Heinroth, Lorenz).

spot under cover and creeping in with a long-drawn softly throbbing call: 'leading a hen to the nest'.

In all such more complex activities, the individual component elements appear in a particular sequence; that does not mean, however, that no element of the sequence can drop out. In fleeing behaviour, all components except the last one can even vanish. This occurs, for example, when the stimulus is sudden and strong (Fig. 8.29b), or when another simultaneously strongly activated activity, for example sitting, prevents all but the last element (Fig. 8.30b). Then

Stimulus

a

• Looking about

• Standing up
• Clucking

• Walking around
• Defecating

• Doubling back Crouching
• Aiming

Flying aw
• Cackling

b

Scream Flying off

Stimulus

Figure 8.29. a: behaviour sequence of fleeing from a ground enemy with slowly increasing stimulation of the field (or constant stimulation of moderate intensity); *b:* the reaction to a sudden, strong stimulus.

the hen behaves like a bird sitting fast on the nest, leaping away only in the last moment of danger.

One can arrange complex behaviour sequences according to various considerations. Whether they are released by the physiological correlate of a sensation, a hallucination, or of a bodily condition (such as hunger or tiredness), is only *one* possible aspect. Another consideration is how far or how completely they are performed without the appropriate external situation. One easily obtains complete fleeing behaviour even in perfectly familiar surroundings. The behaviour of a cock is also complete, when with increasing stimulus he first looks tensely into the distance, then fearfully closer and closer by, and finally leaps away to the opposite side – exactly as if a rat were passing by. In a hen one can obtain from a stimulation field the true-to-life behaviour 'defending chicks against a goshawk' with wing arching, feather erecting, running around in small circles and piercing screeching; living chicks, which the hen in question other-

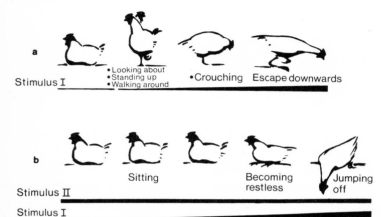

Figure 8.30. a: behaviour sequence of fleeing from an aerial enemy with slowly increasing stimulation of the field (I); *b*: the same with simultaneous stimulation of a second field (II) which produces sitting down.

wise pecks, are peacefully tolerated with even weak activation of this stimulus field.

These are examples of the one extreme; of 'vacuum activities'. Other behaviour may also take place 'in vacuo', but is more intense when the appropriate object is offered; this applies to the 'wing-scratching' (*Kratzfuss*) of a cock (Fig. 8.21) in the absence of a hen. Finally, there are behaviour sequences which we have thus far never

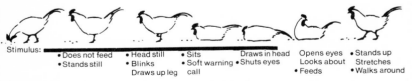

Figure 8.31. Behaviour sequence during a sleep stimulus and (after the end of stimulation) while waking up again. The elements designated by ● are obtained in isolation from other stimulation fields.

observed *in vacuo*, such as the attack of a cock upon a rival or an enemy, or in the hen the feather-plucking and pecking of a socially lower-ranking hen. They require at least a suitable dummy as object, and are performed the more expressively, the better the dummy is. Fig. 8.32a gives as example the attack of a cock on a stuffed polecat. Lacking a polecat, the next best object is the face of the keeper –

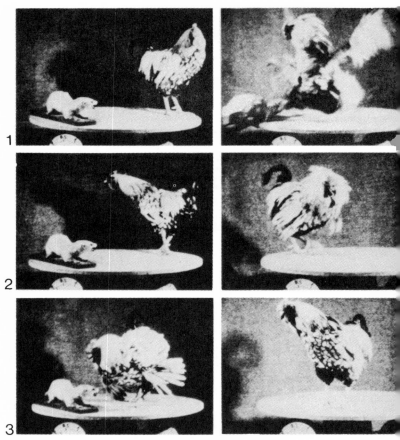

Figure 8.32a. Attack on a stuffed polecat produced by a brain stimulation. (1) Before stimulation the cock stands unconcernedly to one side. (2) Stimulation begins, cock looks attentively towards polecat. (3) Stronger stimulus (0·2 v). Full enraged posture. (4) Directly afterwards, attack with spurs. (5) The enraged mood continues. (6) A triumphant crow following the end of stimulation.

not his hand, for that has hen-characteristics (Fig. 8.32b). Lacking even this object, one sees at most a threat gesture into empty
space.

Now it is most noteworthy that all complex behaviour sequences
are composed of elements which can also be released in *isolation*
from *other* stimulus fields. In Figs 8.28–31 those elements which

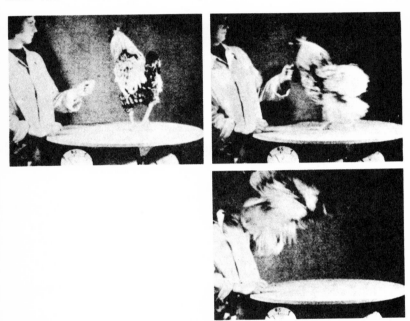

Figure 8.32b. The same as *a*: the hesitant attack is made on the face of the normally
well-liked handler.

could also be obtained in isolation are each designated by a dot. Compared with their appearance in the whole sequence, behaviour elements thus 'cut out' often seem automaton-like; a hen which *only*
clucks takes on the aspect of a clucking machine; the cock who
shows indefinitely the oscillating aiming movements of the head,
as they appear in fleeing only shortly before taking flight, or the
cock who from the whole hunger behaviour only shows uninterrupted snapping and swallowing *in vacuo* – and as if the mouthfuls
got bigger with increasing stimulus strength – all look as if they

251

had been bewitched into senseless activity! This impression is always strong when the uniformity is great, but adaptation and mood shift are slight.

If one collects the elements which can be isolated from all higher behaviour sequences, it appears that many such elements can occur in quite different sequences. Sitting, for example, can be a part of sleeping or of brooding, or on the other hand it can be 'pure' sitting. Clucking and defecating accompany various behaviour patterns which have the character of anxious tension; cackling is a sign of removal of tension after fleeing as well as after laying an egg; looking around, standing up, and running are common to numerous behaviour sequences; standing ducked and motionless can belong to aerial-enemy behaviour (freezing) or it can be a gesture of submission to a stronger rival. Even food-pecking on the ground can still be either an element of hunger behaviour or an element of fighting between rivals, where it appears as a threat gesture ('displacement picking'); we have found stimulation fields for both cases, in the second case a rival fight follows at once when one presents another cock (even a stronger one!).

The fact that particular behaviour elements can belong as components to completely different drives (and moods), is theoretically important. It forces us in every case to search for and to *name the highest integrated unit* activated by the stimulus field. For if the terminology does not reflect the synthetizing capacity of the CNS, the reconstruction of the functional organization is made impossible from the beginning. Let us explain this postulate of *level-adequate terminology* more fully with an example. General motor unrest is a common reaction. Only gradually have we noticed that this unrest generally refers to something quite specific. One must vary the surroundings in every possible way to discover the actual goal of the activity. In the experiment in Fig. 8.33, conspecifics and various other objects were ignored; presenting a fist released slight threatening. A stuffed polecat, fastened to the table, however, *together with* the stimulus, at once produced vigorous threat and attack. Are we then dealing with a tendency to attack an enemy of the species? No, the term is still too specific; for if the stimulus lasts beyond the attack, the hen checks for a moment, turns, and flees screeching. The 'resistance' of the (motionless) enemy, but only *together* with the brain stimulation, results in the reversal into flight. The adequate expression for the field, including all possibilities, must therefore be something like: 'ground-enemy behaviour' – in distinction from other fields, which can produce attack alone or fleeing alone.

The locomotor unrest, which in the example of Fig. 8.33 is the initial element of enemy behaviour, is in another case in the cock directed only towards the head of a conspecific, which is at once

Figure 8.33. Releasing ground-enemy behaviour. Without a suitable object the stimulated hen shows only locomotory unrest. Towards a fist she shows only slight threatening (*a*). A stuffed, motionless polecat is vigorously threatened and attacked. If the stimulus ends at this moment, the hen remains standing and threatening slightly (*b*). If it does not end, she checks and flees, screeching (*c*). (After cine film records.)

pecked (rival fight); in a third case towards water, which is drunk (thirst), or a hiding place, in which the cock calls (leading to nest), and so forth. If one were always content with the diagnosis 'loco-motor drive' and then attempted to study its relationships to other behaviour, the chances of obtaining comprehensible relationships would be slight indeed, since we are in actuality dealing with the most varied goal or appetitive behaviour! The same is true of all movements which sometimes appear 'pure', and at other times as parts of one or another higher drive.

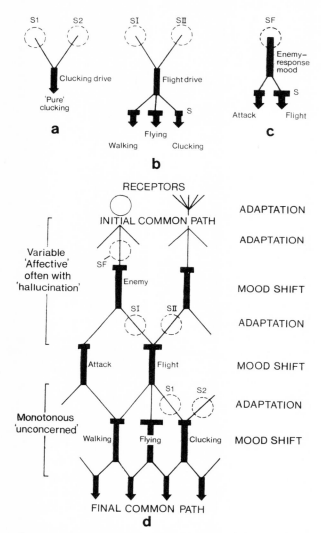

Figure 8.34. Sketch of a fragment from the functional organization ('*Wirkungsgefüge*', control pattern) of some behaviour patterns in the fowl. Explanation in text.

A fragment of the functional organization

Most of the road to the reconstruction of the whole multidimensional functional organization still lies before us. We will content ourselves here with a small segment, in order to show a possible method of proceeding. Our goal is only to arrange some behaviour patterns which contain a common element, clucking. Starting point is the experiment already sketched in Fig. 8.16c, with two stimulation fields which produce 'pure' clucking (Fig. 8.34a, S1, S2). Next to it is shown the result of a second such experiment, which differs from the first only in that here clucking is a part of fleeing behaviour, so that with increasing drive – we simplify somewhat – clucking, walking around, and flying away follow one another (Fig. 8.34b, S1, S2). We interpret this sequence with different thresholds of the individual acts for a common superordinated fleeing drive, expressed in Fig. 8.34b by crossbars (T) of different length. Finally, we take still another experiment, in which, this time from a single stimulus field (Fig. 8.34c, SF), threat and attack appear with weaker, fleeing (with all its components) with stronger central stimulation,[1] which we again explain with different thresholds for the same drive of the behaviour. Thus we have three levels of integration before us, whose arrangement is not open to doubt; Fig. 8.34d[2] shows this arrangement. 'Pure' clucking lies closest to the motor apparatus, the 'final common path' as Sherrington called it. The enemy behaviour lies closest to the sensory apparatus, which we might call the 'initial common path'. For just as a particular muscle is active in various reactions, a sense organ is involved in releasing various behaviour patterns.

The rough sketch (Fig. 8.34d) already shows some interesting things. One sees that on the whole a network results, as the doctrine of plasticity demands, but it is a network of *specific* effects. One recognizes, furthermore, that to all appearances adaptation and mood shift occur repeatedly in different layers of the CNS. And finally it becomes clear that structures closer to the motor apparatus

---

[1] Anthropomorphically interpreted: the dangerousness of the enemy increases with the strength of stimulation.

[2] The stimulus field S1, which produces clucking, is only arbitrarily drawn on the pathway leading from the fleeing drive; it might just as well lie in any other afferent path, as long as it is at the same level. The same is true of SI, since fleeing drive is not activated only by enemies.

initiate stereotyped reactions, whereas structures lying closer to the sensory apparatus release behaviour sequences in which the animal 'as a whole' appears to be involved.

Another look at the localization problem

At the outset we discovered some facts which made the localization of functions in the CNS problematic: (1) a stimulation field appears 'silent' at times, but at other times active; (2) over a period of hours, changing behaviour patterns can be activated from a stimulation field; (3) the same action can often be released from spatially separated fields.

The attentive reader will have remarked that these three phenomena are now physiologically understandable, and in fact necessary.

(1) It is to be expected that stimulus fields remain silent when a dominant behaviour such as freezing or fleeing tendency, which suppresses other activity, is spontaneously (and even slightly) present. To say that the stimulus field is 'silent' is of course incorrect, since the field is indeed excited, but the resulting activity is blockaded somewhere else. (2) To understand the change of reactions from one and the same field, we refer to two phenomena: the invisible propagation of excitation away from the point of stimulation (cf. Fig. 8.16d) and the huge variation in the individual thresholds of behaviour through mood shifts and adaptation, with its consequences for the dominance relationships in conflict situations. If one makes the assumption, which is certainly often correct, that the stimulus field includes structures (for example pathways) which belong to different neuromotor systems, and which are all simultaneously – invisibly! – excited, it then necessarily follows that the decision as to which behaviour must become visible is made not in the stimulus field, but *somewhere* else, by the total dynamic situation prevailing at the moment. (3) That identical reactions are to be obtained from different fields follows in part from the circumstance that so many movements are elements of quite different higher behaviour sequences; one thus expects that pathways will lead to such a neuromotor apparatus from various places. On the other hand, this scatter in localization is certainly in part an artifact, resulting from not adhering to a level-adequate terminology. When, for example, all stimulus areas from which 'locomotor drive' can be activated receive

this label, no matter whether 'pure' motor unrest or thirst-, hunger-, nest-, rivalfighting-, or enemy-behaviour – which cannot become visible through lack of the appropriate external object – is involved, then the scattering of localization *must appear* much *larger* than it actually *is*! In concrete terms, we may designate sitting down, standing up, walking, looking around, clucking, defecating, threatening, and so forth as 'pure', when neither raising the stimulus to several times the threshold value, nor varying all external factors, succeeds in eliciting anything else. Very often, however, the question remains open, for example because of disturbance from neighbouring structures which are also stimulated; such uncertainly defined stimulus areas are therefore useless for localization.

From the foregoing arise automatically certain precautionary rules for future localization attempts; whoever is interested can easily derive them. It is our conviction that the still puzzling relationship between function and structure must become much clearer if these rules are observed.

Conclusion

Perhaps it has become clear to the reader that the construction of a functional scheme of the brain stem activities is indeed a difficult undertaking, but not a hopeless one. Of course monistic doctrines, which try to 'explain' *the* function of the CNS according to one formula, are useless here. All the more, however, are we moving with our thinking and method of expression into the neighbourhood of differentiated human psychology. It is to be hoped that with closer contact human psychology and the physiology of behaviour will learn from one another; the former, what physiological mechanisms must be taken into account with humans too, and the latter, how one obtains a finer-meshed terminology, better adapted to reality. We do not need to fear that in the process 'anthropomorphisms' will sneak into behaviour physiology. It may, however, very well become clear how 'theriomorph' man is.

Translator's note

The English translation of this paper owes its existence to the instigation of Prof. W. H. Thorpe. The translator wishes to express his

257

indebtedness to Dr C. Fraser Rowell who originally undertook the task of translation, and to Prof. von Holst, Dr von Saint Paul and Dr H. Mittelstaedt, who read and approved the manuscript.

9 Tactile illusions in estimation of thickness through touch   *1962*

WRITTEN IN COLLABORATION WITH LISA KÜHME

Our ability to judge the thickness of thin plaques by touching them with the fingers has apparently been little investigated, when seen in the context of this age of measurement technology. This present study was sparked off by a marked illusion which one of us (von Holst) repeatedly experienced in this respect.[1] Some verified results have emerged, and at the same time the study has shown that further experiments would be very profitable.

If experimental subjects are given an aluminium plaque of 5 cm diameter and $2\frac{1}{2}$ mm thickness and told to select a similar plaque from a series graded in intervals of 0·1 mm in thickness, they achieve this with a mean error of 4–10 per cent.[2] The error is approximately of the same magnitude with one-handed touching (in succession, with the hand alternately grasping the test plaque and one in the comparative series) as with two-handed (simultaneous) touching. It is important that the plaques are rounded at the edges and that optical aids are excluded.

The influence of mass

If experimental subjects are instructed to select an aluminium plaque of the same thickness as a plaque of lead, one of aluminium, one of wood and one of cork, the curve shown in Fig. 9.1a is obtained,

---

[1] Whilst working on moulded plaques for violin construction.
[2] A number of extremely uncertain experimental subjects showing wide scatter in judgement were excluded from further investigation.

provided that all of the plaques are suspended on threads. If the test plaques (along with the comparative series of plaques) are stabilized in wooden clamps, the curve is much flatter, but not horizontal. Thus – and this is the first tactile illusion – the moved mass is incorporated in the estimation of thickness. The stabilized cork plaque

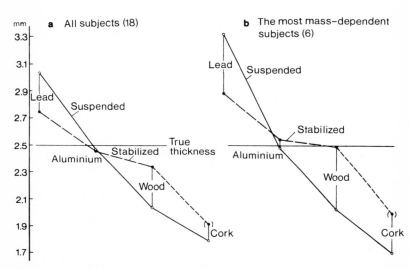

Figure 9.1. Comparison of $2\frac{1}{2}$ mm thick plaques of lead, aluminium, wood and cork with a series of aluminium plaques ranging from 0·9–3·3 mm thickness. (Mass ratio = 30:8:2:1.)

differs from the others, apparently because it remains pliable – which is interpreted as indicating that it is thinner. In Fig. 9.1b, there is a selection of the group of experimental subjects most influenced by mass; the thickness illusion in the case of free suspension shows a 2 : 1 ratio (lead : cork).

The 'corrected' curve with stabilized plaques is, however, no steeper than the corresponding curve for all experimental subjects. In other words, the correction produced by stabilization is greater.

The influence of finger pressure

If a plaque is held by pressing the fingers together with a pressure of 1–2 kg, the terminal digital pads are flattened to such an extent that the nails are brought 1–2 mm closer together. It is to be expected that any possible coarse error in thickness estimation resulting from this would be compensated. Fig. 9.2 shows that in many experimental subjects there is actually *over-compensation*. When two brass plaques of 1 mm thickness are brought into contact against a spring resistance of 1·4 kg and an aluminium plaque of the same thickness is sought with the other hand, it is found that some experimental subjects will select the same thickness as when tested with a simple brass plaque of 2 mm thickness, whilst others will exhibit a

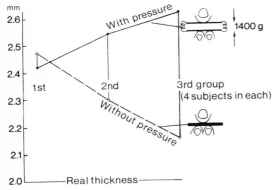

Figure 9.2. Comparison of a simple brass plate, or two brass plates, which can be compressed against a spring resistance (total thickness of 2 mm in both cases), with a series of aluminium plaques of different thicknesses.

small or large difference in favour of the pressurized plaque (groups 1, 2, 3). Remarkably, the pressure *is* actually perceived as a resistance; but the mean errors are no larger than with normal touching. This over-correction is most likely to be attributable to the tendon spindles; but receptors in the terminal digital pad (the least likely being the muscle spindles) might also be involved. The question could doubtless be resolved experimentally. It can also be seen from Fig. 9.2 that the brass plaques are estimated overall as thicker than the aluminium plaques, in the same way as lead (Fig. 9.1). There is a negative correlation between this illusion and the pressure illusions, though this cannot be regarded as generally established in view of the fact that there were only twelve experimental subjects.

The influence of surface contour

A further, extremely conspicuous, illusion arises when (stabilized) curved plaques of aluminium are compared with the flat plaques in the series. Fig. 9.3 shows the values for the experimental subjects who were least (a) and most (b) dependent upon this curvature illusion. This is despite the fact that the plaques were grasped with both fingers (as shown in Fig. 9.3b – above right). The illusion

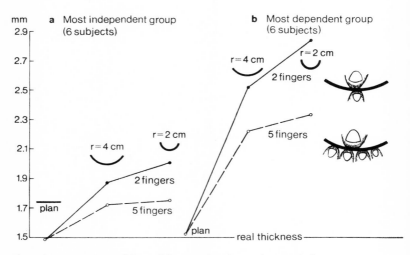

Figure 9.3. Comparison of flat, mildly curved and strongly curved aluminium plaques of 1·5 mm thickness with a series of aluminium plaques of different thicknesses. (——— touching with two fingers; - - - - touching with five fingers.)

increases with the degree of curvature; but there is no increase when sections of spheres are used instead of cylinder sections.

However, as soon as the subject is permitted to touch the curved plaques with all five fingers, as shown in Fig. 9.3, the illusion is considerably reduced (by 30–50 per cent) – in some cases quite abruptly. Thus, it is obviously the additional information provided by the four externally applied fingers which corrects the error, at least to some extent.

Tactile illusions in estimation of thickness through touch

The effects of temperature

The dotted curves in Fig. 9.1 give rise to the suspicion that thermal conduction may also be an operative factor in causing illusions. We therefore carried out a number of experiments with a hollow brass plaque (2·5 mm thickness), through which we could pass water at different temperatures. First of all, we tested nine experimental subjects to determine which plaque of the comparative series appeared to possess the same thickness as the brass plaque at 10°C and 40°C

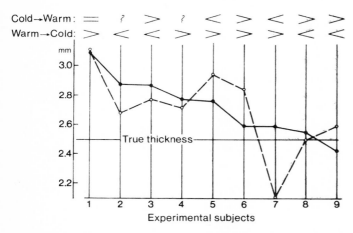

Figure 9.4. Comparison of a 10°C (●) or 40°C (o) brass plaque 2·5 mm thick with a series of aluminium plaques of varying thickness, involving the nine experimental subjects. Signs at the upper margin: alteration of sensation of thickness accompanying abrupt change of the plaque temperature from cold to warm (upper row) or from warm to cold (lower row). < = plaque appears thicker; > = plaque appears thinner.

respectively (Fig. 9.4, 1–9). As can be seen, there is no systematic arrangement in the results; it frequently occurs that the colder plaque appears to be thicker – but it is also common for the warmer plaque to appear thicker. If the experimental subjects are only permitted to touch the plaque whilst the temperature is rapidly altered (in a matter of seconds), there is almost always a sensation of change in thickness, which is often extremely pronounced (Fig. 9.4, upper two rows). Nevertheless, this sensation is not necessarily in harmony with the estimation of thickness with a constant temperature (cf.

experimental subjects 4 and 9). There is therefore much which remains to be explained.

Finally, in two-handed simultaneous experiments, we also varied the temperature of the hand used to touch the test plaque by previous application of warm or cold water. In this situation, there are some experimental subjects who exhibit a sympathetic change in the other hand in line with the temperature change. As far as the rest are concerned, only the following general statement can be made. There is a temperature effect accompanying change in the temperature of the hand; but it is only in some cases that there is a clear-cut

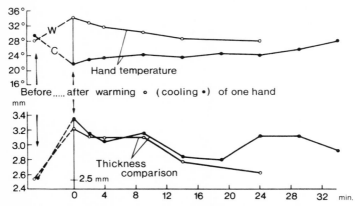

Figure 9.5. Two-handed (simultaneous) comparison of an aluminium plaque of 2·5 mm thickness with a series of aluminium plaques of varying thickness, before and after change in hand temperature (one experimental subject: each point is the mean for three experiments).

pattern. Fig. 9.5 provides an example in which cooling, like warming, produces a marked and regular increase in perceived thickness. There are also experimental subjects who exhibit an irregular effect, only amounting to an overall increase in scatter of the recorded values. Finally, there are also subjects for whom neither warming nor cooling the hand produces any effect. An example of the latter is provided in Fig. 9.6.

The temperature experiments described are incomplete in every direction, and they are mentioned primarily as a spur for further study. Probably, one will only be able to gain some insight into these phenomena when the entire thermal balance of the experimental subjects is controlled before and during the experiment.

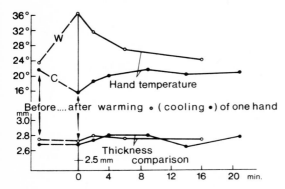

Figure 9.6. Experiment as for Fig. 9.5. One experimental subject: each point is the mean for two experiments.

Summary

When thin plaques of the same material are compared in one-handed and two-handed experiments, the mean error is 4–10 per cent. When plaques are suspended, those made of lead and brass are estimated to be much thicker than they are, whilst wooden plaques and cork plaques are judged as being somewhat thinner and much thinner (respectively) than they actually are. The errors are much smaller when the plaques are fixed in position. The compression of two plaques against a spring resistance leads, with many experimental subjects, to an overestimation of overall thickness. The thickness of curved plaques is considerably overestimated, and this effect increases with curvature. The illusion is very pronounced when touching with two fingers and far less obvious when all fingers of the hand are used in touching.

Changes in the temperature of the plaque and in the temperature of one hand (in two-handed touching tests) both produce clear-cut illusions of thickness in many cases, but so far no clear inherent principle has been identified.

Part Three
Borderline questions
in biological research

10 On the nature of animal life *1947*

Anyone who picks up a modern biological or physiological textbook in order to find a succinct answer to the question as to what actually represents the 'nature of animal life' will doubtless replace the book in disappointment, having realized that science always offers us stones instead of bread. In fact, the question concerning the 'nature' of life is never posed. Instead, the reader finds himself overwhelmed with an enormous quantity of individual facts, relationships and laws which are arranged, somewhat forcibly, under headings such as 'metabolism', 'stimulus phenomena', 'development and growth', etc. It appears as though biologists, who after all should be concerned with the theory of life, have become entangled in an endless wealth of phenomena and have long ago forgotten their original task – namely, the recognition of the actual significance of life in general and of the inherent nature of animal life, as distinct from human life.

If the enquirer were to turn to a living biologist, rather than to an inanimate book, he might well hear the following answer: 'My dear fellow, no field of knowledge can give you what you are looking for. For you, any description of living processes and, indeed, any attempt at conceptual delimitation of living things would appear empty and dead alongside what you yourself imagine or sense behind this word. The kingdom for which you are seeking the key does not belong to our world. The person you *really* need is somebody who could succeed in uncovering the indeterminate feelings and semi-conscious thoughts which emerge within you in association with the word "life" – phenomena which are not (as you believe) conjured up by the Nature which you observe, but produced by the creative force of your own soul. This could perhaps be achieved by a philosopher, certainly by a wise man, and most certainly by an inspired poet. The natural scientist, the biologist is much more modest. He is satisfied with observing and measuring Nature, and establishing in the realm of external perception an organized pattern according to rules of reason. He is indeed acquainted better than others with living organisms and life processes; but he has no knowledge of the "nature" of life itself. For such a man, this word has no scientific sense – only a philosophical, metaphysical one. Since Galileo, more than ten generations ago, declined to ruminate over the nature of gravity and began straightforwardly to follow the movement of falling objects through measurement, thus discovering the gravitational laws, this sober pattern of thought has enriched the human spirit with new insights with every year that has passed. It has eventually provided human beings with power over our environment and ourselves – a power which would be fitted to a more noble humanity.'

This is the kind of answer given by the biologist. But the enquirer will perhaps turn away in boredom. It is also possible – though rarer – that he will begin to list and to pose further questions. In this case, he will gradually realize that our human spirit is at home in many realms, and that it can speak many 'languages', each of which is only meaningful as long as it remains free of any mixture with the others. He will recognize that all of these languages – such as those of science, art, philosophy and religion – occupy particular spheres in human life and that, in consequence, a so-called naturalistic or biological *'Weltanschauung'* ('world view') is nonsensical for the simple reason that the natural sciences of course legitimately occupy only *one* of our spiritual compartments.

It will also be understandable why biology remains dumb when confronted with questions which are not phrased in its own language. If, instead of asking about the nature of animal life, we pose the straightforward, modest question as to which life processes distinguish the remainder of the animal kingdom from the animal species *Homo sapiens*, biology finds its voice, and we can hear roughly the following:

'Nowhere in metabolism, in stimulus phenomena, or in development and growth is there a notable dividing line between man and

270

animal – and the same applies to the structure of the body. It is only when we come to activity and inactivity, to behaviour, that a gulf appears to emerge between man and the other animals, dividing off the "powder baboon" (as Morgenstern called him) as the animal which can tell lies and commit suicide, and which has a history.' (All such definitions are based on man's peculiar behaviour.)

Probably, the most embracing characteristic of man is his history – man is *the living organism which is bound by tradition*. The behaviour of the individual is not set out step by step by his hereditary factors; instead, it is determined in detail by the historic development in which he is incorporated. In different cultures, his thoughts and activities assume different forms. There are rapid changes as a result of historical catastrophes, without any accompanying change in the hereditary complement, which in fact only determines that the individual is, by nature, a creature of thought and learning – a creature of tradition, and thus (at least potentially) a creature of culture.

We human beings tend to regard this as the great *freedom* which characterizes us above all other animals. Seen through the eyes of an animal, this could equally well be termed an unbearable *compulsion*. After all, in the course of a long period of childhood and youth in which there is as yet no possibility for critical rejection, an extensive behavioural code – ranging from language and custom up to life philosophy and religion – is imposed and imprinted upon us! It is, eventually, a matter for individual decision whether a man happily swims along with this stream or desperately attempts to free himself from it. Sluggish tradition operates through the mass of fellow being unremittingly on each individual and holds him capture with iron claws. The animal world is free of such a harness; in the absence of any tradition, and simply following inner drives and inhibitions, animals live out their lives in a paradise of unrestricted activity.

However, such a judgement, made from an animal point of view, would certainly be just as one-sided as our human preoccupation with freedom. In the comparison of man and animal, we are not confronted with *a different degree of one and the same kind of freedom,* but with *different forms of restriction*. We human beings – creatures of habit that we are – are unable to conduct our lives without the numerous aids which we owe to our ancestors. Through learning, we adopt for ourselves what they created, and we use this foundation of tradition to continue the work of constructing our culture.

But how are animals able to live without these aids; how are their environmental, familial and social relationships regulated? One can, in fact, find highly developed animal 'states' with organized division of labour! One thing is certain: the popular comparison between human and animal 'states' falls down in the most important point. The knowledge and ability which every individual human being must learn afresh from outside is, in animals, organized into 'states' (e.g. bees), already anchored in the hereditary complement, and emerges spontaneously from within in the course of a gradual maturation process. For example, each bee, in the course of its life, carries out each of its predetermined professions in turn – care of the larvae, comb-building, food-gathering, etc. Division of labour in the beehive arises exclusively through the age differences between its occupants. These complicated behavioural sequences are by no means rigid as a result, appearing in blind obedience to an internal command. Instead, they can adapt to the varying external situation such that even pronounced disturbances can still be counteracted. For example, if all of the younger comb-building bees are trapped out, there is an initial short period of unrest and apparent 'disconcertion' and then some of the old hive occupants at the stage of food-searching revert to comb-building.

For some time now, we have traditionally referred to the sum of such feats (which often appear to be quite remarkable) by using a well-known word, which – in the role of a refuge for our ignorance, a *Deus ex machina* – provides the appearance of an explanation. This word is *instinct*. In order to say what the biologist thinks of this term 'instinct', one would have to repeat all the things which have already been said about 'life'. The biologist has no knowledge of '*the* instinct' and its 'inner nature'; for him, this is a metaphysical concept about which natural philosophers might speculate. On the other hand, he is acquainted with a wide and varied range of drive-governed (or, if you prefer, *instinctive*) *behaviour patterns* – that is, observable activities – in the animal kingdom. And just as Galileo, unconcerned with the inexplicable nature of gravity, once dropped spheres from the leaning tower of Pisa in order to identify the laws of gravity, so the biologist must set out from simple drive-governed behaviour patterns in search of the difficult laws of animal behaviour.

We can sketch the path of such research with a number of

examples involving animals which possess, like us, family bonds, and in which the difference from the human behavioural mechanism should therefore emerge quite clearly.

How do young animals find out the difference between enemy and friend? Freshly-hatched wild pheasant chicks at once exhibit the correct behaviour when a predator approaches. This is because the adult bird which they are following has two different alarm calls. The first is uttered when a terrestrial predator (e.g. a fox) is spotted, and the young birds respond to this by promptly clambering into the nearest suitable bush. The other alarm call, which the mother utters when a predatory bird approaches, equally reliably induces the chicks to hide and 'freeze' under grass and other low vegetation. If we take the liberty of describing this example with psychological expressions, we would doubtless have to say: The young birds possess 'innate knowledge' of the meaning of these two alarm calls. The fact that something of this kind exists should not really surprise us any more than the well-known fact that we ourselves also exhibit responses without any prior learning, such as closing of the eyelids when a small foreign body is propelled towards the eye, or coughing when dust enters the trachea. In short, we ourselves carry out a number of activities 'as if we know' what their utility is. In our own case, this involves unconscious – more exactly, *involuntary* – action. It is impossible to test whether (and to what extent) the same applies to animals, and whether they in fact possess anything comparable to our consciousness. For this reason 'consciousness' is also a metaphysical concept for the behavioural investigator, and it is therefore left untouched.

On the other hand, one can test how much can actually be 'known' without learning from birth onwards. A Muscovy duck which hears the distress call of a mallard duckling will come over and defend it just as she would with one of her own offspring. If the young mallard is freed, however, the Muscovy duck will at once ill-treat it and drive it away. This behavioural disharmony is explained by the fact that the freeing response of the adult bird is elicited by the duckling's distress call, which has the same sound in both species. Care of the ducklings, by contrast, is elicited by the coloration of the duckling, and this is different in Muscovy and mallard offspring! It would appear that the duck does not possess a conceptual image of the infant as a whole, but that specific characters of the offspring elicit specific activities from the mother. Correct overall behavioural performance is thus based on the fact that the infant combines within itself all the necessary releasing characters. This is the actual situation, and these behavioural releasers which are 'known' from birth onwards are always simple yet characteristic shapes, colours, sounds or movements.

This can be demonstrated by raising an animal in complete isolation, as a 'Kaspar Hauser', so that it never sees its own kind. One can then use simplified imitations (dummies) of conspecifics to test what a releasing substitute image must possess in order to elicit a specific instinctive behaviour pattern.

Young, inexperienced partridges seek cover in the undergrowth when an artificial predatory bird dummy (in the form of a black cross) approaches in the sky, whilst they are not at all affected by a round disc or a rectangle. Young blackbirds and starlings, which normally stretch towards their parents with their mouths gaping when the latter arrive with food, will (if artificially hatched) gape in the same way to a flat, round cardboard disc, as long as this has a small lateral projection or an eccentric spot (marking the head of the parent bird) and as long as it is moved to some extent. The 'knowledge' of greylag goslings about their parents is even more limited. When the young goslings hatch from the egg, the first large moving object they see becomes the parent. Thus, if the human caretaker is alone with them at hatching, they will follow him exclusively from then on, however much the true parents might try to attract them. The goslings are imprinted on the human being.

It would be misplaced to attempt to explain the simplicity of the necessary releasing stimulus in all of these cases in terms of restricted visual powers in the young animals. Instead, it is exclusively an indication of the economy of such genetically transmitted semi-knowledge. In the course of juvenile development, the image of the conspecific typically becomes enriched by increasing addition of experience. This can eventually develop to such an extent that in a colony of gulls, which all seem the same to us, parents and grown offspring can individually recognize one another among thousands of others.

Many peculiarities in shape and colour are, accordingly, explained through the fact that they serve as innate releasing stimuli for particular behaviour patterns. In the spring, our bullfinch dances around his chosen mate with a small twig in his beak, and – by stretching himself and spreading his feathers – he makes his breast as long and broad as he possibly can. The more luminous red he presents in this way to his loved one, the greater is the likelihood that she will take notice of him. Such 'demonstrative behaviour' is widely distributed among animal suitors, and anyone acquainted with this is struck with the number of resemblances to our own youthful male behaviour.

If we turn our attention away from the stimulus releasing the behaviour to the *drive-governed behaviour pattern* itself, the latter also proves to be an innate behavioural element on closer examination. The view that birds learn to fly and that their nest-building is improved through practice has proved to be unfounded. The appearance of additional learning is often a misleading product of internal maturation process. A young pigeon whose wings are temporarily bound down will exhibit just the same improvement in flight after a few days as its siblings, which were able to fly unhindered in the meantime. Animals *do not learn* their activities in the same way that we human beings must learn to speak and to utilize the body, particularly the hands. When animals learn something, it involves the *application* of their innate ability to one object or another. In this respect, their nature has in many instances left room for individual experiences. For example, ravens initially collect all kinds of possible (and impossible) objects as nest-material and attempt to fixate them in the twig framework with stereotyped pushing movements. Only gradually do they learn that twigs always stay attached as a result, whilst chippings repeatedly fall away. Eventually, they will only collect utilizable material, leaving anything which is unsuitable.

It has long been recognized that one feature of such drive-governed behaviour patterns is the vehemence with which they strive for emergence. One might believe that this is based on the insight which the animal has into the utility of its actions (either for the animal itself, or for its progeny). But this is not the case; instead, the impulse is towards performance of these behaviour patterns for their own sake. Just as a young man in love does not recognize that his activity is geared towards preservation of the human species, a dog set loose in a wood does not hunt to avoid starving – initially, at least, the dog hunts in order to hunt. Tame antelopes in enclosures where they are never exposed to any danger will now and then storm off without any external provocation, simply in order to give vent to their dammed-up escape drive. A starling which is well supplied with food catches 'imaginary flies' in a room definitely free of flies and 'swallows them' with enjoyment. Thus, the drive-governed behaviour pattern has an emotional foundation. When its normal elicitation is absent, it will discharge itself in a vacuum in its usual form, thereby demonstrating that a need is present, whilst consciousness of the goal is lacking. This also explains why animals kept in isolation literally fall in love with their keeper, or with his hand. As Lorenz says, 'with this potion in their bodies, they see Helen in every woman'[1] – though in this case the potion is nothing more than the bitter medicine of enforced abstention.

In this way, animals are reliably conducted through their lives by a neatly balanced system of innate releasing stimuli, behavioural acts and interposed learning dispositions. This equilibrium is only disrupted in cases where man intervenes – with domesticated animals. In such cases, under pressure of selection exclusively for human benefit (that is, usually in the direction of fattening, docility and reproductive drive), we find deformities and omission effects. Our domestic hen, which happily cackles beside a freshly laid egg, offers an example of such omission. Its wild predecessor – the Burmese jungle fowl – initially slinks away after egg-laying and performs its cackling (i.e. the alarm call against terrestrial predators) at a *different* site, thus attracting any predators away from the nest. With the domestic hen, this behavioural sequence has been rendered senseless by the loss of one of the links (that of slinking away).

Indiscriminate and exaggerated expression of the sexual drive is also a typical product of domestication. With wild forms, there is an introductory ceremony involving innate behaviour patterns (often of a complex kind) and sometimes a long period of betrothal prior to the final act which Schopenhauer called the 'fireworks'. It should be said in defence of natural Nature that – both in the farmyard and with human beings themselves – the various deviations which we are accustomed to call 'animal' or 'bestial' should more correctly be called 'human . . . all too human'.

In the end, we thus encounter the disquieting suspicion that we civilized human beings are also in the process of falling victim to our self-domestication. Do we not find – quite apart from the touchy subject of sexual partnership – troublesome omissions of *innate* gov-

[1] A quotation from Goethe.

erning principles in various other fields, such as in social cooperation? In the wolf pack, where there are (as everywhere else) rank
fights, the defeated partner presents his most vulnerable aspect (the
neck with its carotid artery) to his opponent – and the victor immediately releases the vanquished. In mediaeval tournaments between
knights, it was similarly regarded as a noble act for the victor to
bid the loser rise and to help him bind his wounds. What has happened to such behaviour nowadays in our age of human mass aggregation? Is it still education or already degraded nature which determines that such behaviour is, at the most, only to be found regularly
on the sports ground? The question must remain open; here, it is
no longer the zoologist but the sociologist who is qualified to speak.
If the latter's decision should one day confirm our suspicion, then
of course no amount of calling 'Back to Nature' will restore what
has already been lost from our hereditary possessions. On the contrary, this exhortation would doubtless be misunderstood as implying
the above-described sense of 'bestiality' (i.e. domesticated degrada-
tion), which is exactly what must be eliminated. Therefore, *reason*
would have to find another way to replace what has been lost and
to divert us away from the chaos of human social decay through
provision of *traditional* governing principles of gradually increasing
stability and general applicability. This would, obviously, gradually
widen the gulf between animals and men and separate us further
and further from the animal paradise of innocence and unrepen-
tance. The naturally desired and unthinkingly achieved interaction
of individuals would be replaced (both in the minor and in the major
contexts) by an *all-pervading political sphere*.

The above attempt to arrive ultimately at a new view of the peculiarity of human nature, through more profound understanding of
the various forms of animal behaviour, derives from a still adolescent
branch of biology. The names of the pioneers in this new field –
that of comparative behavioural research – are Oskar Heinroth and
Konrad Lorenz. It is hoped that this essay may at the same time
represent a note of remembrance for Heinroth, who was sadly taken
from us at the end of the war, and for Lorenz, who was taken as
a Russian prisoner of war whilst serving as a doctor.

11 The human environment and technology *1952*

The naïve human being lives in a colourful world of objects and phenomena, of joy, pain, memory and hope – and all this variety is bound together through its relationship to an unchanging ego. Before philosophical or scientific doubts begin to trouble us, everything which the ego experiences represents 'reality' in a uniform sense. However, since about the time of Galileo, science has divided this naïve world into an objective, measurable, generally accessible – in short, 'material' – part and a non-objective, non-spatial, non-measurable 'spiritual' part which is private to each individual. In the course of a long process of purification, the latter has been excluded from physics. This separation into two worlds was not at all easy, and it was often achieved in opposition to the naïve evidence of our own experience. The fact that, for example, the vividness of different colours does not correspond to anything which occurs in the objective world and that it is a product of the perceiving soul – whilst physics simply distinguishes wavelengths which can be investigated equally well by a colour-blind physicist as by one with normal colour vision – has, like many other facts, only slowly gained acceptance. This process of purification ultimately – through the general theory of relativity – even affected our concepts of space and time. Just as Kant taught that these categories are preconditions for any objective perception, properties which stem from our minds and not from the spatial environment itself, physics excluded perceived space and perceived time from its range of competence. What remained was a system of mathematical relationships of a complex nature, which are quite obscure and are only really accessible to a small number of specially structured physicist brains. One might regard the whole thing as a refined 'game with glass beads', were it not for the fact that it involves a conceptual system which has enabled man to achieve, through the intermediary of technology, unexpected dominance over Nature. If we take the word 'reality' in its literal sense, as indicating something which takes effect upon us, then abstract physics is also, in its own way, extensively real.

But what becomes of the naïve, colourful, perceived world in whose exclusive fold we are able to move with familiarity – the human *environment*. Who can explain its origin, after the physicist has renounced any further interest in this direction? Here, the materialist would answer: The human, perceived world is nothing other than brain physics! We must find methods of ever-increasing refinement for investigation of the brain, and then we will encounter the realm of spiritual experience. It is easy to recognize that this view is absurd. What is the brain? An organ composed of thousands of millions of ganglion cells, which look like spiders with legs of various lengths. If these legs were joined up to form a single thread, they would stretch from the earth to the moon. It is thus an extremely complex organ within which equally complex processes are operating at all times (in sleep as well as during activity) – processes which we can to some extent follow by recording of electrical events. If one wished to give a detailed account of the actual events in the human brain occurring during the space of one minute, listing of the electrical processes alone would itself fill an entire library. For these processes take place with tremendous rapidity. A single ganglion cell can transmit several hundred impulses in the course of a second. We are, without a doubt, only just beginning to become acquainted with these processes. But we can already say the following with certainty. Wherever research may lead us, we will never touch upon spiritual, upon experiential events. The argument of the materialist can most easily be extrapolated *ad absurdum* if it is taken quite literally. It would mean that one could construct an apparatus which I could attach to my neighbour's brain such that if he, for example, were suffering from toothache, I would be able to perceive his toothache (his spiritual experience) through the apparatus: *I* would have a toothache in *his* tooth. As you can see, we can of course obtain quite exact knowledge about the physical processes which take place in another's brain when he is experiencing pain, but pain itself remains the private possession of that individual.

Such an example clearly shows us that we actually do live in two worlds and that only one of them is fundamentally open to inclusion in the realm of the natural sciences. We do, in fact, know that the other world (that of experience) is correlated with the first, in some paradoxical manner, but that we are unable to directly grasp this world itself. We can, however, do one thing. We can study more closely the material conditions operating in human beings which, in some sense, run parallel to what is experienced and which, canalized by external influences, also exert an obligatory modifying effect upon experience. In this way, one can learn more about the environment, and such knowledge is assisted by medicine, psychology, physiology and also by zoology, the study of the behaviour of organisms. I should now like to take you on a short excursion through this scientific field, which is still in its infancy. It will be seen that the problem of human beings and technology presents itself here in a novel and extremely remarkable manner.

It has already been mentioned that everything which I experience takes place in time, and that much – namely, anything outside me – occurs in a spatial context. We shall now take a closer look at perceived time and perceived space and examine the conditions which are associated with these experiences. If I should ask my neighbour, 'What is time, really?', he would probably pull out his watch and show the hand which rotates and measures this parameter. This is no answer to my question, but it is an indication of a physical means of apprehending this remarkable phenomenon – time. And we can at once ask, 'Does our organism incorporate anything in the nature of a clock?' The answer must be 'Yes'. We do not, in fact, know exactly where this clock is located; but we can already state that it is not a physically operated clock – it is a chemical device. There is a general rule that chemical processes are temperature-dependent, accelerating as the temperature increases. Patients suffering from malaria, which leads to rapidly alternating increases and decreases in temperature, have been requested to make time estimations by marking an estimated time-lapse with a movement or a sound at certain time intervals. It emerged that these time-lapses, which were experienced as equally long, were objectively slower in passage at low temperatures and passed more rapidly at high temperatures – in agreement with the above-mentioned chemical rule.

This internal clock is something which has been considerably neglected in civilized society. But there are animals in which the clock operates with great precision. My friend and colleague Gustav Kramer has discovered in our institute that one can condition birds to run or fly in a specific compass direction. For this orientation response, they make use of sunlight, which they can also recognize through a cloud layer. Just as we can determine a compass direction by using the position of the sun and a pocket-watch, a central apparatus within the bird can perform the same process. The bird's clock (whose location we have not yet discovered) thus indicates the exact time of the day.

I have been speaking above about physiological *measurement* of time in human beings and animals, and this must, of course, be distinguished from time *perception* itself – the experiential aspect which we shall now consider. It will soon be evident that perceived time may have no strict relationship to the physiological clock. In the first place, our experienced time is divided in a quite peculiar manner into past, present and future. The present seems to be something like the point of contact between past and future; but closer examination shows that the present has a given duration, even if this is difficult to determine. In order to experience the present, one must – remarkably – possess a memory. You cannot grasp the sense of the sentence you are reading as a whole if you have no trace of a memory and therefore no connection between the first and last word. Anybody who temporarily loses his memory completely as a result of an accident also loses his perception of time, and this loss takes with it any understanding. Even consciousness of the 'ego', which of course binds everything together, seems to have disappeared completely. *Experienced* time is, in fact, something which can exhibit extraordinary variations in objective rapidity. It can even happen that we simultaneously experience different phenomena with differing speeds. I can, for example, read a book out loud, whilst at the same time my thoughts are darting about and performing a long chain of events. This need not necessarily prevent me from following the meaning of the text which I am slowly and loudly reading. The reader has doubtless already encountered situations of a similar kind.

The most amazing increases in rapidity of experience occur in certain dreams. A large number of well-documented cases have clearly demonstrated that dream experiences can be tremendously swift. It can happen that somebody falls out of bed and at the same time dreams a long story of a mountain climb terminated by a fall from a rock; at the moment where the dream fall occurs, the subject lands on the floor and awakes. It has been the tendency to state that anything of this kind is physiologically impossible; yet this is a completely erroneous view. I have already said that our brain processes take place extremely rapidly. Within a single second, a tremendous variety of organized excitatory processes can occur in the human brain, and I am therefore convinced that one should turn the problem around and ask: 'Why is it that in the normal course of waking experience we think so *slowly*, and that what happens in our consciousness does not occur at a consistently greater velocity?' And my answer to this question is: 'This is correlated with the fact that we human beings are organisms which can speak.' Speech is, after all, a physical act; the mouth must be moved, air vibrations must be produced, and this must all be done at a speed which is far from arbitrary. But if I want to say something coherent, a thought process with the same speed must be performed in order to direct this motor act of speech. In adaptation to human speech, slow thought processes have been developed at quite a late stage in phylogeny – and this is the form which we civilized human beings regard as 'usual'. But alongside this, pushed far into the background, there is an evolutionary vestige from the early age of thought through rapid sequences of illustrative images and the like, rather than through speech concepts. And this earlier form of thought can take place far more rapidly. A number of arguments can be advanced to support this interpretation. Disruptions in the left temporal lobe of the brain lead to emergence of a pathological condition referred to as 'motor aphasia'. The patient is no longer capable of word formation, and he can no longer speak in an articulated fashion. He also lacks speech concepts; but thought in terms of illustrative images is retained. Thus, it can be seen that these two things are separable. Or, alternatively, one can say: 'There are certain chemical influences exerted upon the brain which may block articulated thought and all motor expression of speech whilst experience in the form of illustrative images is at the same time enormously increased.' This is the effect of opium. Under the narcotic influence of opium, an individual seems to be sitting in a stupor when observed from the outside; in his internal world, however, there is every indication that he experiences a variegated and colourful realm of indescribable beauty. He quite literally lives in one night through an entire, colourful life of a hundred years. Apparently, this experience has a beauty and plenitude which vastly overshadows that of our everyday existence. It is understandable, in view of this, that anybody who has once experienced this will wish to return again and again. Thus, we can see here for the first time that there are agents which can direct our experience – in this case our experience of time.

With this, we shall leave the experience of time (an area which is still packed with unexplained problems) and turn to our experience of space. The space in which we live is also segmented. In contrast to time, however, it is divided into *above* and *below* and into *near* and *far*. Once again, we can ask: 'Which mechanisms in the human organism ensure that this segmentation is so arranged?' With respect to the distinction between above and below, a naïve observer might think that this is a matter for experience. We know that the sky is above and the earth is below, or we know that houses always have vertical walls, so with the aid of our eyes we can always determine where above and below are located. But such an answer would be erroneous. In a room kept in complete darkness, one can arrange a luminous paper strip suspended on the wall so that it is exactly vertical. Indeed, this can be achieved not only when the head is held upright but also when the head is inclined to the side to any given extent. Thus, even without visual aids, we know where above and below are. In fact, we know this as a result of the activity of a small organ located behind the mastoid bone on the left and right sides of the head. This is the *statolith apparatus* – a small, horizontally located stone on either side, which rests on sensitive sensory hairs immersed in a fluid. When the head is inclined, these sensory hairs are bent by the weight of the stone, and the bending produces nerve impulses which are conducted to the brain, where – in the literal sense of the word – the size and direction of the inclination is 'calculated' from the information input. Of course, this calculating apparatus can only function correctly when it receives correct information from the organ concerned. For example, when the statolith apparatus is put out of action or disrupted on one side, there must be a persistent erroneous input of information about the position of the head, and this is actually the case. The result is that any animal or human being afflicted in this way has the persistent, erroneous impression that the world is tipped to one side. Balancing movements are carried out to correct this, and the body falls over. This small calculating apparatus can also be led astray in a more elegant fashion, by electrical stimulation of the nervous pathways which conduct the impulses to the body equilibrium centre (as has been done by our colleague Baron Romberg). In order to do this, it is simply necessary to attach electrodes to both mastoid bones and to pass an extremely weak oscillating alternating current through them. The result is quite peculiar. Firstly, the limbs and eyes perform restricted equilibrating movements. Secondly – and this is far more important – when one opens one's eyes and looks at the room (whilst the weak stimuli, which are not actually perceived as such, are passing through the nerves), the room is seen to rock to and fro rhythmically. The artificially deceived equilibrium centre, which cannot know that an electrical apparatus is producing these impulses, must conclude that the head is rocking, and the result is that it must conclude that the immobile environment is similarly oscillating. Once again it can be seen that a small apparatus determines what we should call above and below, and this apparatus can be technically influenced.

But an organ of this kind is not exclusive to man; the same thing is found in all vertebrates. However, it is not the only organ found in Nature which tells the organism where above and below are. In fish, for example, above/below is additionally determined by the eyes. If I should remove both statoliths (which is quite easy to do), above and below are exclusively determined by the direction of incident light. If I illuminate an aquarium from below, I can induce such operated fish to turn upside down and swim with the back pointing downwards – for these fish, 'above' is always the direction from which light arrives. Under free-living, natural conditions, above is always indicated by the momentary location of the sun. Thus, such animals have a heliocentric environmental image, whilst we human beings have only an earth-bound (geocentric) one, as a result of our postural orientation apparatus. This is a small physiological difference, yet it has great historical consequence. You are aware that our entire civilization, including (for example) Christian civilization, is geocentric, and you also know of the great struggle which the human intellect has had to endure, since Copernicus, in order to establish the fact that the earth is moving and that it is the sun which is standing still. All of this is a result of the fact that we have a geocentric postural organ. If we were built like fish, Copernicus would never have been necessary, since nobody would have been able to sell us the tale that the sun moves and the earth stands still. So much for spatial segmentation into 'above' and 'below'.

Space is also characterized by its *depth*. Some things are close to me, while others are far away. If I were to take a photograph of the objects in a room, I would be able to tell from their size on the photographic plate how far away they were, provided that I have prior knowledge of their real size. The rear wall of the human eye – the *retina* – resembles a photographic plate in that objects are portrayed as images whose size decreases with increase in distance from the eye. But what we actually experience is quite different. I do not perceive a given object, such as a cane or a stool, as being smaller, the further away it is. Instead, I perceive that it always has the same size, but is at some distance from me. Once again, a small refined calculating apparatus is involved (one which has recently been discovered), and it is this which ensures this perceptual feat. Let us carry out an experiment which anyone can easily repeat. A bright cross is visually fixated for some time in a dark room. As is well known, this produces a so-called 'after-image', i.e. after a while, the experimenter can turn his eyes away and yet still see the bright cross, which has been (so to speak) clearly engraved on the retina. If an after-image of a distant, bright cross is produced on the retina in this way, and the gaze is then turned to a nearby surface, the cross is perceived as lying on this surface; but it has changed its size to become remarkably small. If the surface is moved further away,. the cross increases in size again. How is this possible? After all, I know that the cross engraved on my retina does not change in size. This phenomenon comes about in the following manner: When my gaze is directed at objects close by, a specific motor command (a nerve impulse) produces a greater curvature in the lens of my eye, so that the nearby objects are clearly focussed. Remarkably, this command turns around and returns like a boomerang to my cerebral cortex, once again penetrating my consciousness. But this time it acts as a sensory illusion – it provides the information that the observed object is smaller. In the experiment outlined above, in which a cross is engraved on the retina such that nothing alters on the retina, the sensory illusion is directly experienced when the gaze is adjusted for close-up vision. But if the gaze is directed at a genuine cross in the distance, and this cross is then approached to the eye, there are two simultaneous events. Firstly, the closer the cross comes, the greater is the reduction in size of the cross exerted by the above-described sensory illusion upon perception. But, at the same time, the cross portrayed on the retina – just like an object approaching a photographic plate – increases in size. The illusion which gives the appearance that the cross is becoming smaller, and the illusion which gives the appearance that the cross is increasing in size, cancel one another out. As a result, perception concludes: the approaching object is maintaining a constant size. (This is, of course, objectively correct.)

One can also influence the above-mentioned, small apparatus by external means, e.g. chemically. A specific poison – atropine – when dripped into the eye, will paralyse movement of the eye lens such that close-up adjustment is impossible. If this is done, even when the attempt is made to look at objects close by, distant objects continue to be clearly focussed on the retina, and in principle the product is the same as with the engraved after-image of the cross. And the effect permits one to predict that the closer one attempts to bring the gaze, the smaller the observed landscape will become. If I were to look into a normal room with an 'atropinized eye' and attempt to look at an object close by, the only thing that would happen is that the room would shrink to the size of a room in a tiny doll's house. This is yet another instance in which my experience can be directed by influencing a specific apparatus in my body.

In many animals, spatial dimensions are assessed in the same way as in man; but this does not apply to them all. For example, bats do not use their eyes at all; instead they use their ears in a remarkable manner. Bats are almost blind, yet – as has been known for some time – they can reliably find their way in complete darkness. This is achieved by virtue of the fact that flying bats utter rapid, rhythmic sequences of sounds which are inaudible to us, and these sounds are reflected from objects in the surroundings. This echo (more exactly, the special pattern of the echo) is perceived through their highly sensitive ears just as finely as our eyes perceive the surroundings at night when they are illuminated with a torch and light is reflected back to the eyes. With this 'acoustic pocket-torch' (if I may use the term), bats orient themselves with such precision that when they are flying in a strange room they can perceive a tiny nail in the wall and suspend themselves from it. So much for spatial distinction into above, below and depth.

Now the space which we perceive is never empty; it is always filled with objects – objects which are usually coloured. Even the quality of colour has been known for some time to be a special product of an apparatus which is located in the retina and is associated with sensory cells of particular form – the so-called *cones*. These cones are only present in a central zone of the eye. The periphery of the retina is not able to distinguish colours, only brightness and darkness. If somebody should slowly move a coloured pencil towards my eye from one side, I would be able to see quite early on that it is a pencil, but I could only state its colour when it is at close quarters. In addition, these cones are less sensitive to light than the other visual elements – the so-called rods. As a result, the objects we see lose their colour as the sun goes down, but their form persists until some time afterwards. This is why 'all cats are grey at night'. But even during the daytime, when the world is seen in colour, experience of colour is by no means exactly correlated with the wavelengths encountering the eyes. The optical apparatus which records colours operates according to its own internal governing principles. For example, if I enter a room containing coloured objects illuminated with a red lamp, then I initially see things only in a red-tinted light. After a short time, however, my eyes become adapted and recognize the real colour of the objects, subtracting (so to speak) the red originating from the light source and superimposed upon the other colours. Or, if I gaze at a red colour spot for a long time and then look at a white wall, I temporarily see a clear-cut green spot. And if I gaze at a blue spot for some time and then look at a white wall, I temporarily see a yellow spot. This is the well-known phenomenon of *complementary colours*: red and green, blue and yellow are apparently indicated by the same cones. Departure from the normal condition in one direction will, for example, indicate green, and in the other direction it indicates red. If there has been a 'pull' in one direction and this is then 'released', there will be an automatic, predominant swing to red. As you all know, there are quite a few human beings whose eyes possess an incomplete form of this apparatus, so that they are (for example) blind to red and green. The colour experience itself is also modulated by the brain, and it can also be influenced at that level. Certain chemicals, such as strychnine administered in small doses, produce an overall marked increase in colour sensitivity. Other chemicals, such as santonin, have the effect that everything is displaced in one direction, in this case towards violet. Once again, this means that it is possible to influence my experience by affecting central mechanisms.

If we now take a look at animals, it emerges that only the monkeys possess colour vision corresponding to that in human beings, whilst most other mammals are completely colour-blind. They do not possess cones at all, and conditioning experiments show that they do not distinguish colours. Still more interesting is the fact that there are animals which can in fact see colours, but respond to a spectral range different from our own. A classical example of this is provided by the bees, which do not respond to our red as a colour, yet perceive ultra-violet (which has no effect upon our eyes) as a colour quality and can distinguish this from the other colours.

But let us return to human beings. Objects are not only seen to possess colour; they are also seen as possessing a certain form, a certain structure. This strikes us as so commonplace that in the course of everyday life we do not even notice that we dispose of a capacity to construct, subconsciously, a structure from a small range of insignificant data, even in cases where there are scarcely any indications of a form. It can be demonstrated that we possess a quite general 'Gestalt pressure', a tendency to isolate figures from a neutral background. This function is apparently localized in a specific area of the brain, in the occipital lobes; as is indicated by cases in which this area is destroyed. When this happens, there are deficiency symptoms which are referred to collectively as 'soul blindness'. Anyone suffering from 'soul blindness' is not blind in the literal sense; he sees everything and responds to optical stimuli, but he cannot recognize objects. What he sees is a turmoil, a kaleidoscope of colours and contours – he lacks the ability to isolate individual features as a figure from a neutral background. Thus, he also lacks the ability to recognize the faces of acquaintances. Such a deficiency clearly shows that there is in fact a special function involved in the recognition of objects as being the same, even though every object appears different from every aspect and under different kinds of illumination. Only 'Gestalt pressure' ensures that this vital function is maintained. At the same time, it makes possible the development of an artistic trend in which the figurative element is increasingly simplified, in which there is increasing abstraction from object content. I am, of course, referring to artistic representation, particularly drawing and especially that of the cartoonist. When an able caricaturist draws four or five lines on a board and all the onlookers identify the figure with one accord, it is this 'Gestalt pressure' which ensures that, despite the fact that such lines have virtually no physical relationship to the face of the person concerned, a clear-cut experience is called to mind.

In this way, arbitrary forms are accentuated as such in our perception. But there is yet another special capacity which accords distinctive evaluation to quite *specific* forms and makes them more important to us than all others. We innately recognize these forms as special and we respond to them appropriately without any prior learning being necessary. I shall illustrate this emphasis of specific forms with respect to animals first, since the situation appears to be simpler. For example, gallinaceous birds know from birth onwards the form of a silhouette of a predatory bird in the sky, something which may mean danger. Chicks can be reared from the egg and kept under conditions such that they will never see predatory birds, and they can then be tested with dummies cut out of cardboard and passed over their heads on a wire in order to find out what shape the contour must be for the chicks (which have never been able to learn from experience) to exhibit the correct flight response. Such tests show that the frightening dummy must possess specific shape characteristics in order to be correctly recognized. A black disc has no effect, a cross has a slight frightening effect, and if the wings are pointed and the head is rounded, the dummy is completely effective. One can also find innate recognition of particular patterns in the auditory realm. For example, the pheasant has two different alarm calls. The first is uttered when a terrestrial predator approaches; the second is given on approach of an aerial predator – and the pheasant chicks respond correctly to these alarm calls even if they have just hatched from the egg. When the terrestrial predator alarm call is uttered, they attempt to climb up into the undergrowth, and when the aerial alarm call is heard they immediately crouch beneath tufts of grass.

Numerous further examples of such 'innate knowledge' (Lorenz) can be found in the animal world. In fact, exactly corresponding behaviour is found in human beings as well. The suckling infant can already see from the facial expression of the mother (and from her vocalizations) whether she is amicable or annoyed and will respond with the appropriate mood. Nobody has ever had to learn the correct interpretation of facial expressions; we all recognize them innately. To take another example. Living organisms which are soft and rounded and have large heads, large eyes and short limbs (e.g. rabbits, robins or even something like a doll) innately arouse within us tender sympathetic feelings and often evoke a tendency to pick them up and cuddle them. Such animals, or objects, elicit the response which is actually tailored for young children. The latter also have large eyes, a large head, short limbs and a certain softness, and these characters of 'lovability' are shared by infants with a large number of dummy objects. It is entirely because of this that there is a doll industry, and this also explains the fact that in the German language the diminutive suffix '*-chen*' is added to the names of animals in this category, such as Kanin*chen* (rabbit), Rotkehl*chen* (robin) and *Küken* (chick); but not to similar, non-emotive animals, such as Hase (hare) and the naked, 'ugly' offspring of sparrows.

We can take this a step further, since this emphasized evaluation also applies to higher, specific forms of activity, to typical behavioural acts (Lorenz). If I see that a child is being mistreated by an adult, then I become uncontrollably annoyed with that adult, even if the entire scene is a dummy performance – for example a film passage played by bad actors. When there is a fight between two people and the victor tortures or kills the vanquished person, I unhesitatingly take sides against the stronger of the two, even if I am only reading an account in a cheap novel. If, on the other hand, the victor spares the loser, or even helps him to his feet, then I experience an uplifting feeling of satisfaction. There is a large number of such typical human behaviour patterns which produce a quite specific emotional effect in the observer, whether he wants this or not. It is a vital part of the skill of novelists and film producers to know how such emotions can be reliably elicited, since the mass success of their products is based entirely on this – and the political demagogue uses the same technique.

In all such cases, we are concerned with quite elementary, entirely innate human behaviour patterns which are equivalent to comparable activities in animals. When a number of fish of a particular kind (*Pterophyllum*) are closely observed in an aquarium, it soon emerges that they have established a hierarchy among themselves. There is a strongest individual which has power over all the others, and a weakest which can be molested by the others whenever they wish; the others are arranged on a ladder between the two. However, it is very satisfying to see how the strongest fish maintains a watch to ensure that the others do not continually attack the weakest, regularly making punitive attacks as soon as one of the middle-ranking fish begins to mistreat a lower-ranking individual. Or, to take another example; in all animals which are equipped for offensive behaviour and live in social groups, there are specific behaviour patterns which prevent the strongest from killing the others one after the other. In wolf packs, bipartite fights are common; but they never finish with the death of the loser. As soon as the fight, which is carried out with full muscular force, leads to submission of one of the opponents, the latter presents to the victor his most vulnerable aspect – the neck with its exposed carotid artery – and the stronger animal immediately refrains from biting. The weak animal is not killed; but from that moment on he is dependent upon the victor. Quite similar behaviour, which we regard as thoroughly 'respectable', is found in many animals. It is only animals that are not adapted for offence (e.g. doves) which, when forcibly confined together, will gradually kill one another by persistent, mild pecking and continuous further penetration into the wounds produced. This is because they lack such inhibitions. Any animal which is endowed with powerful weapons by Nature is at the same time presented with the appropriate 'ethical code' – innate 'knowledge' of the situations during fights with conspecifics where the weapons may no longer be utilized.

The reader will understand that the obvious step from here is to consider man and his technology. Nature has scarcely equipped human beings with weapons of destruction. Only modern technology has provided us with the possibility of killing other people in large numbers. But such technology was unable to equip us with the appropriate ethical inhibitions. Imagine, for example, that a bomber pilot were to receive an order to go into an air-raid shelter and to strangle with his own hands all the women, children and old people that he would normally wipe out *en masse* by simply pressing a button and allowing his bombs to fall without any special inhibition. It is obvious that the pilot would be quite unable to carry out such killing with his own hands, unless he were a pathological criminal. Thus, it can be seen how technology has utterly disrupted the balance between weapons and the natural feeling for the limits of weapon use, and how our lives today suffer from the fact that we are searching for generally valid higher ethical forces which *Nature* did not *necessarily* have to give us.

With this, I come to the end of my excursion through the peculiarities of the human perceived environment. We have seen in various ways that it is possible to influence our perceptual experience by influencing the mechanisms in our bodies upon which such experience depends. And here a new form of human dominance over human beings emerges. In fact, we are already acquainted with several kinds of dominance. One is that of direct force; this has always existed and nowadays it is used in a massive way, e.g. in the form of abduction. The second, which makes use of refined technological methods, is dominance through the psychological entrance; that is, through influence of human attitudes via a number of channels (the press, radio, cinema, etc.). This is a process which does not usually penetrate to the conscious level during our daily lives, and which is therefore difficult to escape – particularly for somebody who is socially attuned and has a need for contact. It is easier for human beings who are antisocial by nature; a fact which itself throws particular light on the problem of so-called 'swimming with the stream'.

In addition, there is a still developing form of influence upon human beings exerted by technical means – that of influence through physiology. Pharmacological agents and surgical interventions, as we have seen, can directly affect my experience, and nobody can escape such effects. The ego itself is altered in such cases. We know that such influences are often employed in a beneficial manner in the field of medicine, such as sleeping pills, pain-killing agents, or even surgical operations on the brain which remove or divorce unbearable pain or unbearable experiences from the afflicted person.

But all of this is no more than a beginning, and advances in research will lead to increasing differentiation, so that sooner or later it will be fundamentally possible to steer human motivation, voluntary behaviour and attitudes to one kind of behaviour or another. The beginnings of influences of this kind already exist here and there, and one does not need much imagination to picture what this can lead to in the future.

Here, as everywhere, new knowledge is accompanied by new power, which can be used for good or for bad – and here, as everywhere, it is necessary to distinguish quite clearly at an early stage what is good and what is bad. To express this quite baldly: Should human beings continue to work, fight and die for what is right and satisfying (that is, for ideals, which one may transiently approach, but never quite attain); or should we follow the simpler pathway – the physiological way – which can, through a small surgical intervention or a specific drug, provide us with the reliable experience of ultimate satisfaction, so that some planning clique can steer and direct a malleable population without application of force, without wars and atom bombs? This question can only be answered by somebody who has some idea of the ultimate purpose of humanity on Earth. I can only pose the question: every man must answer it for himself.

12 Belief, power and the physical concept of the world   *1958*

In history lessons, one learns of times past that when two peoples with different religions waged war upon one another, it was quite common for the vanquished people to adopt the victor's religion – not through compulsion, but quite spontaneously because the victor's religion had proved to be the *true* one owing to its efficacy. A similar interpretation was involved in mediaeval divine judgements, where the victory of the stronger party signified divine agreement that his opinion was the correct one. Napoleon's statement 'God is always on the side of the strongest battalion' is no more than a cynical inversion of a fundamental religious feeling of this type, which is also reflected in innumerable conquests by miracle; for miracles can only be achieved by somebody who has power over *all* natural processes.

Thus, whether we like it or not, experience of power and belief are closely associated in the human soul; it is scarcely to be avoided that belief becomes established wherever power is experienced. In the early family fold, the father is the physically embodied principle of power, and thus childish belief experiences in 'our dear God' the paternal attributes of leadership, rectitude and goodness. But, for most of us, these times are long since past. If we look around nowadays for impressive power, our gaze is irresistibly attracted to technology. Whilst the churches have the same 'to offer' as one hundred years ago, every year sees the appearance of new and more admirable technological power, whether in the form of electronic brains which can perform in one minute as much as a human calculator in the course of his entire life, or in the form of atomic weapons competing with one another for destructive capacity. The mind of the layman, which means the minds of most of us, experiences something of this kind – implicitly or explicitly – as a *miracle*. And since, as everybody knows, technology is only the organ of execution – whilst the spirit which liberates or confines such forces resides in the brains of natural science, of physics – consequently, physicists unsuspectingly appear to be endowed with priestly abilities and graces, of which there would not be so much as a trace if their professional activities did not extend beyond their four walls. In fact, physicists are not only surrounded by this aura in secret; they are even asked directly and publicly to answer questions of the following kind: Is religion possible? Can God intervene in world events? Does free will exist? And – almost unbelievably – many of the recipients of such questions actually hold forth and, drawing upon their better *physical* knowledge, give the 'true answer' to a respectful and credulous audience.

One must ask how the general confusion of concepts could extend so far. How could we arrive at such disregard of the limits of cognition which were once very wisely established by the natural scientists themselves and to which they entirely owe their proud edifice? Let us attempt to explain this.

As is generally recognized, natural science is based upon phenomena occurring in the external world, on things and processes which are fundamentally accessible to anyone, which are measurable, and whose laws can be elucidated through experiment. It is not the internal *nature* of the attraction between masses but only the *measurable process* which should concern us – so runs the thesis of Galileo and Newton. And ever since, this kind of conceptual approach has characterized the course of this science. Even in cases where living organisms were involved, this method was extremely successful. Anything which could be objectively demonstrated in their lives and activities was investigated; but what they *experience* in the process (or whether they actually do possess experience) remained untouched, since it was not testable.

On this extremely confined but solid foundation the skyscraper of natural science was erected; the only intellectual edifice (apart from the mathematical one) in which anybody can feel at home, whatever the colour of the skin beneath which his brain is housed and whatever the language in which his thoughts proceed. For all

295

subjective experience, everything which adds colour to our existence, and everything which binds men together or distinguishes them internally, remained rigidly excluded from this realm.

As this proud edifice grew and developed its first world-wide effects, its influence gradually spread through the minds of men. The feeling that the material world of physics is the *actual* reality, and that all phenomena obey natural laws, gradually expanded. Spiritual values and religious content faded away. In brief: the materialist philosophy began to spread, replacing the immediate reality of our experience with the impressively effective (and thus presumably *more truthful*) conceptual framework of a science. Instead of leaving the *physical concept of the world* on its confined site in the kingdom of intellectual existence, where it belongs, it was made absolute and, through inversion, a small place was made *within it* for the accommodation of all intellectual and spiritual phenomena (the human brain). The materialist believed that, with refined apparatus, every feeling and every thought would be uncovered *in reality* as physical processes within the brain.

Let us stay with this important point for a moment. The fact that experience is closely tied up with brain function is a matter of everyday experience; coffee and alcohol, pain-killing tablets and brain damage all bear witness to this. But what is implied, for example, by the following statement: My experience, such as toothache, is only *apparent*; *in reality* it is a physical process in the brain. Can it in fact be anything other (and actually *more real!*) than exactly this and this alone: my persistent toothache? A train of thought must be quite distorted when it involves eradication of a direct factor with words! What general uprooting of the soul must result if it falls prey to such dialectics! In concrete terms, what must be the minimal operation of an apparatus which would render my toothache (and nothing other than my pain) the object of scientific research? If it were attached to my head, for example, it would have to produce the effect that a certain Professor Müller or Dr Schulze could have a toothache – a toothache in *my* tooth and not in theirs! (For then it would not be my pain, but their pain – i.e. something uncontrollably different.) But such absurdity does not shake established belief; quite the contrary. The materialist *is* a believer; he unerringly believes in the reality of his thought-spectre.

Nevertheless, there remained one area in which attempted absolutism in terms of the physical world was not really successful and where the church, in particular, was able to provide successful opposition: the field of causality. If everything, including my actions, is causally determined, even free will would be no more than an illusion, something which does not exist at all *in reality*! It was inappropriate to admit this, simply because of the practical consequences themselves. In line with such reasoning, one would not be able to punish a criminal, whose action was causally and thus compulsively predetermined. Of course, this train of thought is also utter nonsense; for if the criminal acts under natural compulsion, the same must obviously apply to the judge who makes judgement upon him – and to me, as I say this – so that everything would be back where it started. But more important than such criticism is the feeling (after all, I can at any time do something or leave it) which really determined widespread resistance to the thesis of all-pervading causal predetermination. People were readily prepared to accept that animals act under compulsion; but man is different – he alone is free!

However, at the beginning of this century, quantum physics emerged, and with it came recognition of the fact that at the lowest dimensions of physical processes rigid predetermination is replaced by mere statistical probability. Reliable predictions for individual cases are fundamentally impossible in this situation. In short, the operational quantum (to use the formulation of some physicists) is equally *free* in its action. What could be simpler than to transpose this *freedom* to the human brain as well, regarding it as parallel or even equivalent to free will. Thus, the experience of free will was no longer *apparent* – it was *physically real* – and the materialist philosophy was accordingly given a new support in its most sensitive region. But on this basis (believed to be scientific, yet actually ideological) the physicist could decide upon ethics and religion and even take up a positive attitude to these matters. After all, who would wish to rule out the possibility that God might slip back into the human brain through the (admittedly minute) back door of quantum leaps, if physics itself expressly permitted this? Clever presentation has produced wide acceptance of this new form of the old materialism, since it is purported to be an opponent of the disease which it in fact embodies. One can even hear the guardians of belief enthusiastically agreeing to such 'liberating' activity. The fact that religion might be forging its own shackles for all eternity by drawing upon

297

physical possibility for its statements is by and large inconsiderable. Anyone who permits something today can, of course, forbid it again tomorrow.

However, to be fair, not all the blame should be attached to today's ideological physicists and to their followers – who are particularly numerous in the enclave of the literature. After all, they are only continuing what their predecessors began. As soon as one attaches validity to any scientific data in a realm which has nothing to do with this science, and as soon as one admits that the processes discovered by the physicist in my brain may be significant for the reality of my experience, for the truth of my thoughts, then the first decisive downhill step has been taken. With the advance of physics, the primal phenomenon of free will is first regarded – unjustifiably – as a mere *apparition,* and then – equally unjustifiably – it is declared to be a *physical* reality. It remains to be seen what free will might be taken to represent in the future. In brief: Anyone who does not grasp the fact that natural science is blind to qualities, values, decisions and all spiritual existence, is already on the way to carrying out the process of devaluation and uprooting which characterizes materialist thought and which often initially produces simple material belief in progress, but usually becomes subsequently transformed into nihilism and *Lebensangst* (fear of life).

It is possible to test, as a factual enquiry, whether quantum processes operate within our brains as directors of activity; but all data so far collected speak against this. Voluntary movements are directed by extensive cell systems in the brain. Under certain circumstances, these systems can be electrically stimulated in order to produce physical evocation of one movement or another, and it is not uncommon that such movements are simultaneously experienced as a spontaneous act of the ego. Thus, it cannot be excluded that man will one day be able to physically control the activity of an experimental subject, whilst the latter simultaneously experiences his activity as freely desired. Both statements – the objective and the subjective – would be true *in their own fields,* and they are not necessarily mutually exclusive. They lie in *different strata of human existence.* We are not rationally equipped for distinguishing how these strata are correlated. Only somebody disinclined to modestly admit this could be tempted to describe two such infinitely distinct things, the freedom of human decision and the chance nature of a microscopic process – with the same word, and ultimately to regard them as equivalent.

The ideological disease whose character and origin I have attempted to outline here cannot be healed through mere rejection of pseudo-scientific trespassing. As I explained at the beginning, the disease begins with an influence upon the mind – it has a religious coloration. This anxiety-flavoured, credulous respect for the scientific/technological spirit and for its priests and prophets will only subside when the present storm of technological development has passed, when much of it has become an accepted part of everyday life, and when it is possible for anyone to critically survey the range of applicability of all this *progress*. In short, we must wait until we have passed beyond the present stage of general semi-development. But – even if humanity manages to survive this dangerous period – it will continue to be true that the human mind will be compelled by the experience of power whenever it takes root in belief. The source from which new power could emerge to take such effect is as yet unknown – we must patiently wait and see.

13 Problems of modern research into instinct 1961

It is said that every profession has its failing; the failing for the practitioner of science is his refusal to draw conclusions for human existence from the knowledge which he gathers. This criticism is probably apt; yet, on the other hand, there are many examples of the involvement of scientists in the examination of problems of human existence. Perhaps I will provide a further example today; for only the first half of my lecture is concerned with research into animal instinct – in the second half I shall take the plunge and observe our human situation from my particular viewpoint. Only the first half will be purely factual, whilst the second will largely involve personal opinion; so I am quite prepared for objections and even for frank rejection.

Any behaviour pattern which is driven by an internal impulse as well as being pulled by external stimuli (by what is perceived) is referred to as an 'instinct'. Hunger, thirst and fatigue – along with 'falling-in-love', self-assertion and the impulse to aid weaker conspecifics – are all widely distributed drives in Nature, and each corresponds to specific releasing situations. Even a quite inexperienced bird will touch with its beak anything that glitters if it is thirsty, and this leads the bird to water. In most cases, very simple environmental characteristics which are innately recognized by the perceiving organism evoke a given instinctive activity. Such characteristics are investigated with systematic imitations or 'dummies'. A black cardboard cross moved through the air will provoke a domestic cock to utter the 'aerial warning call'; in the spring, a mere bundle of red feathers will induce the male robin to launch a vigorous attack, as if it were faced with a real opponent. The fact that an animal which is otherwise educable apparently does not recognize an inanimate dummy for what it is, rapidly becomes less striking when one thinks that saccharine inescapably tastes the same to us as nutritive sugar and that a picture of a 'sex-kitten' will in certain circumstances evoke quite specific feelings, although it is actually a simple combination of paper and colour.

Specific relationships exist between the internal drive and the external stimulus. The stronger the drive, the lesser is the minimal stimulus required; and the greater the frequency with which a certain activity must appear during life, the greater is its drive component and thus the ease with which the internal drive alone will evoke the activity. A wolf will run agitatedly back and forth in his cage even when satiated, since it is part of his make-up to cover long distances in the pack, to follow tracks and to prey upon game. The lion, on the other hand, lies down contentedly and will prosper even in the smallest circus cage; for here is an animal which lies in wait and pounces upon game at the water-hole, rather than tracking its prey.

The most interesting drive systems from the human point of view are found in animal species which live as social communities which have a rigid, species-specific structure. A collection of altruistic drives is, after all, a precondition for this, and these drives conflict with the egoistic ones which are always present. A hungry cock which finds an earthworm can either eat it or attract the hens with his feeding summons call. As a rule, he will do the latter – at least, more frequently than a human being would hand in unopened a purse found on a deserted street.

An observer who is acquainted with the entire drive system of an animal species after many years of study is comparable to somebody who understands other human beings. He can (with statistical accuracy) predict how an individual of this species will behave in a given situation. If he is personally acquainted with the particular individual, the prediction will be even more reliable for this specific case. Prediction signifies intellectual mastery, and in the case of human understanding of other human beings or of animals, it has an intuitive character – i.e. one is not completely conscious of the reasons for a given judgement. Such mastery is, therefore, no more than an initial step towards knowledge; the subconscious information must be made conscious. In this, the primary aid is provided by methods of technical mastery of behaviour, in which the animal *spontaneously* does what the research worker requires. This possibility already exists, and is today widely exploited; it consists in direct electrical stimulation of the structures in the brain which, in the natural performance of instinctive behaviour, are excited by specific external stimuli and by internal requirements of the body. These stimulation fields are located in the so-called brain stem, an ancient brain region which has a similar structure in man and all other vertebrates. Using fine stimulation electrodes, such a brain region is sought out and stimulated with weak currents (with a potential of less than 1 volt), whilst the animal is left unrestricted.

In this way, we have studied the instinctive behaviour of the chicken, and we have been able to activate almost all of the drives in its rich behavioural repertoire: hunger and thirst, tiredness and wakefulness, and the tendencies to preen, to seek out ectoparasites, to sit down, to stand up, to strive upwards or downwards, to fight with a conspecific over the rank order, to peck at a subordinate individual, to fight off a predator, to flee from a predator. In the cock, we have also activated triumphal crowing, uttering of an alarm call for an aerial predator or that for a terrestrial predator, courtship of the hen, the search for a nest site and display of this to the hen. Every specific stimulation field in the brain governs a specific drive. If the normal reference object is lacking (e.g. the hen for courtship, or the predator for defence) then a substitute object such as the human hand or some other dummy will be accepted if the current is increased to some extent. And if the drive is activated very strongly, the behaviour pattern will be performed '*in vacuo*' even without any appropriate external situation. The cock will then give an alarm call for a predator which does not exist, give the summoning call for an invisible morsel of food, and court an empty space; whilst the hen will lay a non-existent egg with all the signs of exertion. It is difficult to avoid the impression that the non-existent reference object must be present as a free hallucination.

One can learn more about the internal dynamics of drives by simultaneously activating *two* different stimulus fields and thus artificially producing conflicts between drives. It is here that the finer quantitative interplay of excitatory processes becomes significant. As a rule, the momentarily more strongly activated brain field will suppress the other; but if two drives are balanced against one another, experimental neuroses and the like can emerge. A chicken which is simultaneously provoked to attack and to flee will do neither the one nor the other; it will ruffle its feathers, dance agitatedly up and down on the spot and screech loudly.

Examination of drives with this technique reveals a picture of the natural dynamics of drives. In the normal course of life, there are always several drives of varying intensity acting upon the animal and imposing their requirements. What the animal actually does is always a result of this internal interaction on the one hand and the prevailing external situation on the other. Such study of drives is not without interest for human psychology and medicine; it provides us with simple 'models'. At the present time, we are combining such experiments with the operation of various 'psycho-pharmacological agents', which can produce tiredness, courage, sexual activity, indifference or freedom from pain. The effects seem to be quite comparable in chickens, and the combination of chemical and electrical stimulation permits certain conclusions about the manner in which such substances affect the brain stem.

At this point, I should like to break off the description and make a number of observations about *man* and his drives in our present-day environment. It is said that man is a rational creature capable of learning, and this is correct. Human beings have a very long period of youth accompanied by a great impulse to learn, and also – as an essential counterpart – a long period of adulthood with a great impulse to communicate. Thus, a great deal of experience can be passed on; the carrier of such experience is acquired speech. Animal 'languages' are innate and only transmit motivation. Human language presupposes representation of the perceived environment within the brain. The word 'wolf' evokes a physical image of this animal in the mind of the listener. It is in this way that language makes the environment amenable; the human being can deal with it internally rather than with his hands. Such intellectual combination provided the basis for mastery of Nature, science and technology.

Therefore, at first sight our existence really does seem to be governed by rational forces. But this is mere illusion. *In addition to this,* we are still drive-bound creatures just like animals; we possess a drive repertoire adapted to an extremely *ancient* mode of existence in which human beings lived in small clans and had to exist in a struggle against a still untamed environment, against dangerous animals and against other, rival clans. This drive repertoire is not so easy to decipher in human beings as in animals, since a drive can be followed in various different ways. Man is, after all, more versatile in motor terms than most other animals – he can walk, crawl, climb, leap, swim and manipulate; and, over and above this, he possesses a special drive for intellectual motivation and moulding of all his actions.

Let us take a closer look at some of these ancient human drives. We can recognize them because, just as in the animal kingdom, they throng for active release. Like the wolf, man still possesses from his early history a strong drive for movement which nowadays, as part of our 'sedentary' mode of life, is very sensibly satiated as a vacuum activity in sporting activity. Sport would not exist if our ancestors had been ambush-predators. This is a simple, unproblematic example; the *social* drive repertoire has a greater range of repercussions. Man is, perhaps more than any other animal species, a social organism. This is evident even from his sensory equipment. On the one hand, he is a visual animal; yet he also has exceedingly good hearing. Auditory perception of very fine vocal differences, and its counterpart – the great individual variation in voice – are extremely ancient adaptations for social life. They permit personal recognition of individuals from their voices both in the thicket and at night. In a society, it is often more important to know *who* says something, rather than *what* he says.

As a social being, man (just like social animals) possesses a drive to 'move upwards', which leads to competitive struggles, rivalry of all kinds, and ascent of the stronger and the more intelligent in the rank order. He also possesses, like animals, the natural complement to this – the drive for faithful obeisance to the leader. The tendency towards 'blind' following of commands from above is a strong one; this had to be the case originally, when it was only possible to counter natural catastrophes and animal predators by common activity controlled by *one* will. The extent of such kindling of common activity was initially limited and corresponded to the range of a single voice.

Problems of modern research into instinct

Human beings – those creatures of instruction and learning, so versatile in their physical endowment – exhibit, as is well known, a social structure which can vary according to time, place and conditions, in contrast to all animals. *Only* man can form *tradition*, break with tradition and re-create tradition. Special drives which do not occur in the animal kingdom are therefore necessary for stabilization of the *prevailing* form of social existence. One such stabilizer is the drive to regard the momentarily prevalent form of organization as untouchable, as 'holy', as an 'eternal' power to which everyone should uncritically subject himself. This impulse goes so far that injustice *from above* is permitted and even called for. In our religion, for example, 'tests of faith' (i.e. undeserved suffering) and 'grace' (i.e. undeserved happiness) are cornerstones of faithful acceptance. A just God in a rational, juridical sense would not be *above* us; he would be predictable and therefore open to (intellectual) domination. A second stabilizing drive of a more commonplace kind is that for the well-known phenomenon of 'gossip' – continuous, observational interest in the activities of one's neighbour, and the wall of rejection which unbearably isolates anyone who does not obey what is valid *hic et nunc*, such as our divinely dictated monogamy or the divinely dictated polygamy observed by Mormons (with their hands on the same bible).

These two drives, which are known to be extremely virulent, discriminate against anyone who thinks differently – once a member of a foreign clan, later the heathen, today the capitalist, communist, idealist or materialist – in all cases an evil to be suppressed. Together with the drive to rise up as a united group in 'holy' enthusiasm against evil, this produces a really explosive drive repertoire, evolved over thousands of generations long before the beginning of science and technology.

In terms of our drives, we are simply not adapted for our modern, technological world, and this is easy to see. Wherever in Nature a slight degree of insecurity in movement can endanger life (e.g. in crossing a deep crevice on a spanning tree trunk), pronounced dizziness forces the insecure animal to crawl forwards with its belly pressed firmly to the substrate. However, dangerous insecurity at the wheel of a car encounters no drive whatsoever; astronomical accident figures show that mere insight is a dubious substitute. Or: The order to strangle defenceless children, women and old people in an enemy country could not possibly be followed by any psychologically normal soldier. Just as with social animals, in human beings a powerful, drive-bound killing inhibition safeguards the life of a weaker conspecific. Yet the same soldier, when trained as a bomber pilot, will follow orders and press a button, knowingly wiping out thousands of weaker conspecifics. Why can he do this? The drive-bound killing inhibition is restricted to *direct physical* killing and does not apply to button-pressing. In this case, too, insight into the relationship is no substitute for a drive.

And further: Technology permits ever-increasing population density in our living space. Apart from this, human beings usually occur in 'bundles' (Schwabe) – in a skyscraper, in the cinema, in a football stadium or on a camping site. In prosperous years, overpopulation also occurs with animal societies, e.g. in rodents. At such times, one can observe increased restlessness and hasty activity among them, together with increasing aggressiveness against conspecifics, which can extend as far as cannibalism; then there is an outbreak of some kind of plague, and the former equilibrium is restored (F. Frank). Whatever the drive basis of this phenomenon may be, human beings would appear to provide no exception. Medical technology has indeed eliminated plagues from human society; but hasty activity, nervousness and reciprocal irritability have persisted as continuing evils which one attempts painstakingly to suppress. The fact that man is, *by nature*, the *friend* of man is first experienced with surprise at times of great need, for example after a night of bombing – the human imitation of original natural catastrophes.

Yet again: Ancient tradition was attached to the *spoken* word; if the young wished to become important, they were obliged to listen to their elders and treat them with great respect. The spread of literacy has loosened this old attachment, and one can now comfortably read everything which is worth knowing in books. Over and above this, rapid technical progress has ensured that the elder members of society are frequently ignorant of what is important nowadays. The need for respect before the wisdom of old age remains unfulfilled or becomes satiated in another direction (e.g. by worship of the printed word). The old, in their turn, have lost their function and attempt to appear young at all costs (unpublished information from Prof. G. Steiner, Heidelberg).

There is also the fact that the press and radio eliminate the boundary between the word and the voice. The small band of similarly motivated members characteristic of the past swells to an avalanche of people bound by oath to *one* holy flag and prepared to undertake anything. In the words of Jacob Burckhardt: when the Jews came to the oft-praised land of Canaan, they had Jehovah order the extermination of the Canaanites. Nowadays, it is possible to inflame one entire continent against another, in the name of justice, peace, freedom, democracy, future human happiness or of some other holy ideology.

New technology modifies the *forms* of mass orientation. The thirties presented us with radio and the loudspeaker; mass meetings and mass manipulation through crass acoustical suggestion followed. The great era of the shouting orators has now passed. Television favours theatrically schooled conversationalists, who are perhaps in many cases mere puppets responding to strings pulled behind them (whether they are aware of this or not). But these are simply changes in procedure; the principle of human manipulation remains and continues to be fortified. There is no possibility for judging on the basis of *personal* experience about any occurrence outside the most private aspects of our lives. The why and wherefore of all information passes through technical channels whose directors can predict our responses in advance. We are living in the epoch of the Manager.

You have perhaps spotted where these comments will lead. Our existence in the Age of Technology is endangered not because we are creatures of reason, but because we are, in addition to this, subject to quite ancient drives. The ever-growing tentacle of technology has the effect firstly that a human being can kill (himself and) his conspecifics without any elicitation of his 'humanity', i.e. the innate killing inhibition; and secondly it increases almost beyond all limits the susceptibility of the individual and the mass to manipulation. Both effects *de-stabilize* all forms of communal existence and produce a persistent danger to humanity from humanity. An individual will readily admit in calm conversation that war is an outdated and senseless activity. But all of these individuals will still make war when 'bundled' together, and they will want to make war in the future. Possession of a common ideal for which one will fight and die, destruction of evil, united obedience to a higher command, proving oneself as a leader or a follower in the face of danger — all of these noble, primaeval drives deeply rooted in the human psyche *strive* for satisfaction and can therefore be attached quite easily to any flag.

In times of peace or in peaceful countries, these drives will partially appear as typical vacuum activities. Gangs of young people (in many cases with spotless reputations) assemble and carry out courageous mis-deeds or even criminal acts, which are actually quite dangerous and call for vigorous participation – this is the problem of teenage mobs. It would be naïve to lecture such teenagers; one would have just as little success in persuading a lover to fall out of love, or vice versa. Enthusiasm and daring can be neither dictated nor prohibited; they are only activated by direct display or drastic description of conditions which naturally evoke these particular drives. It is apparently not possible to 'abreact' drives of this kind in a passive way in a cinema; man is not built that way – he is a creature of action.

Three partially overlapping procedures present themselves for countering this dilemma: education, suppression and diversion. I do not intend to talk about education, since everybody discusses this anyway. The fact that education, and particularly imprinting early in development, can achieve much is obvious; but it is mainly the form rather than the content of our behaviour which is modified. Just as each human being acquires his own, particular speech word for word, but only does this because he is a creature of speech *by nature* (possessing a drive for such activity, which would compel him to invent language if he were not offered one of some kind), so it is with the rest of our behaviour. The particular forms are plastic, but the fundamental tendencies with which we are exclusively concerned here are amazingly powerful. The stubborn opinion that man arrives in the world as a 'clean page' and can become anything through education is, admittedly, a component of widespread modern ideologies (which are otherwise in conflict); but it is factually utterly untenable.

The second procedure, that of suppression, has a long natural history. Fear is a drive which under natural conditions – and also in our stimulus experiments – will keep other drives in check even in small doses. Once upon a time, man must have remained hidden in his cave from powerful predators or natural forces. Nowadays, such forces can be replaced by a state authority, which spreads fear and insecurity within its realm of power. A man then becomes modest and retiring and only at home can he feel partially protected; he will only be led into daring acts by extreme pressure. This procedure is still successfully utilized here and there, as is well known;

but only a misanthropist would maintain that this is the preferable means of 'satisfying' humanity. In passing, it should be mentioned that long-lasting suppression can ultimately lead to atrophy of natural social behavioural tendencies. This is a highly significant phenomenon which has scarcely been investigated.

An intermediate position between suppression and diversion is occupied by physical and chemical methods, with which man can affect his own brain and those of other people. In the first place, there are psychopharmacological agents. These also have a long history; but their vast general significance is only now becoming apparent in gradual stages. In some countries, the populations of large towns are already involved in a dependence relationship with certain tranquillizing, activating or sleep-inducing drugs which can only be described as a vice. In this field of possible massive profit, industry makes every effort to find active agents of ever-increasing specificity, which activate emotional states or behavioural tendency with ever greater precision. This in fact increases the applicability of the procedure. It should be more widely known than is the case at present that chemical manipulation of human groups is already practised, for example in the army ('passion-killing' tablets; substances which suppress the sexual and rivalry drives of barrack inmates). The dangers of this equally comfortable and unnatural method cannot as yet be foreseen.

The same basic effect (though physiologically more natural) could be achieved through electrical brain stimulation. In actual fact, critical observers have taken the experiments which I have reported as a foundation for speculation about future 'remote control of the masses' by means of this process. I do not believe that this danger exists. It is indeed technically possible to achieve 'remote control' of organisms even without wiring, using a transmitter and a small receiver linked to the electrodes. But all that is achieved in this way is that *one* particular behavioural state, one behavioural goal, is ensured with each stimulating electrode. And even if the operation is somewhat insignificant and entirely painless, what state authority would be able to implant an entire battery of electrodes in the brains of each of its subjects? Those in the know would in any case vigorously oppose this, since the thought of doing with 'free will' something which is actually desired by somebody else is unbearable to anyone. The process will perhaps play a part in medicine: the emotional state of depressives will be uplifted by appropriate stimulation and sufferers from insomnia will be able to fall asleep when required through stimulation of the sleep centre. It is already known (W. R. Hess) that such a form of sleep induction is more natural than sleep provocation through barbiturates.

It is indeed true that pleasurable feelings and perhaps liberation from desires *could* be artificially produced. American research workers have taught rats with electrodes in their brain stems to switch on the stimulus current themselves by applying pressure on a switch with the foot. It emerged that there are brain zones which the rat will stimulate without cessation. We do not know *what* the rat experiences; but it must be something extremely agreeable, since the animal becomes obsessive, missing its food, water and toilet activities, and eventually succumbing if the experiment is not halted. It would certainly be a comfortable and easy method of obtaining pleasure; but only psychopaths will want to use it. It is utterly utopian to think that one might thus satisfy all humanity, that at any selected moment the 'pleasure centres' would be stimulated as a result of personal decision or outside intervention.

Far more serious attention must be given to the highly significant methods of *natural diversion* of our internal drive requirements. I have already mentioned sport as a good means of satisfying the general locomotor drive. But many forms of sport are far more than this: rank-order contests, leadership and obedience, vigorous communal activity (even involving danger to life), enthusiastic onslaught against an 'enemy' – all of this, framed within rigid rules of play, can be satisfied in sport. Where such contests, dramatized by the press and radio, are raised to the level of national political events, their positive significance cannot be overestimated, even if (in objective terms) they are pure vacuum behaviour. The search for further kinds of 'sport', with a maximum of international participation and inherent danger (such as flights to the moon, manned satellites and the like), represents an important task for the future. Much can be achieved in this way; but definitely not *everything*. The insatiable thirst for power, possession and position will continue to be a source of infinite internal friction in everyday human social existence.

Anybody who is optimistically inclined – even the property of seeing the world in 'black' or 'rosy' terms is drive-bound, and rhythmic alternation between the two is characteristic of the compulsive malady of the manic depressive – will be of the opinion that a sensible combination of all the procedures described would be able to keep this internal social friction to a minimum. He might well be right. The optimist is also likely to think that external friction in the co-existence of different peoples could also be reduced; for general peace would at last appear if there were *just one* holy dogma maintained throughout the world. Any basis for enmity and war would then disappear. But this optimistic opinion is *definitely* erroneous. Anyone who thinks this way has still not grasped the decisive point; namely, that there is *not the slightest objective foundation* for our fear-driven, staggering course along the road to mutual extermination. The momentary 'reasons' are, after all, exclusively *produced by man himself* and skilfully camouflaged as factual evidence, *so that* he can follow his primaeval nature and thus eventually strike out. Stone-age man may have had a good, factual reason for chasing neighbouring clans from his own, bare living territory, inspired by 'divine anger' and using all available means of force (i.e. with threatening speech and stone projectiles). Nowadays, it would be sufficient to ensure moderated birth control to limit the continuous, imbalanced increase in the human population, and there would not be the slightest reason to prevent complete harmonious existence everywhere, with the rich excess potential devoted to the wondrous works of art and culture – which at present are senselessly destroyed in wave after wave of combat. Thus, even if a single holy dogma, for example the communist dogma, were to really cover the earth one day, it would immediately disintegrate into at least two, vigorously opposed formulations (one true and the other heretical), and enmity and battle would flower as before – because humanity is, unfortunately, what it is.

In view of this locked situation, it seems to me that the most likely perceptive politician of the future would be a man educated in mathematics and psychology, who maintains a secret agreement with his professional colleagues/opponents over the whole world and steers this agile, capricious and explosive human mass in an organized manner. He would provide diversion here, satisfaction there, and perhaps even set off or tolerate small wars and revolutions in order to avoid large ones. There would be a secret 'team' which would continuously direct fear and contentment, enthusiasm and disabusement, oriented aggressivity and its satisfaction in appropriate doses with all of the available technological channels. The individuals in this well-controlled material, whilst subjectively fully retaining their personal freedom, would achieve no more than a scattering effect around the pre-arranged mean values with which such a world government might operate. Perhaps we are already on the way to this.

As a biologist who is accustomed to look away from the present state of affairs and to follow the modification of organisms in evolution, I cannot fail to pose the following question in conclusion: 'Must humanity continue in the future as it is now – a muddle of the most wonderful and the most destructive natural properties?' Through planned selection, man has adapted domestic animals to their modified living conditions. Disruptive properties and drives have been *bred* out, and desirable features have been exaggerated. In many species, flight and escape drives have been eradicated, whilst in the fighting cock aggression has been enormously increased. To the evaluating observer, some of these changes appear to be *losses*, as in the goose, where the drive for life-long mate fidelity has been replaced by unselective sexual activity without individual bonding (K. Lorenz). But other features appear to be gains; the noble thoroughbred horse is also a pure product of breeding. Human beings have not conducted such domestication with themselves, and this is why man's primaeval drive nature is still present. In civilized existence, there is only the fact that natural selection mechanisms have been practically excluded, and this leads to increasing individual variability.

Any suggestion for planned human selection has become suspect because of misuse. But anybody who has experienced more than one world war, who has children and sees that the future can only bring something worse, if everything remains the same; anybody who is himself a philanthropist and does not share the common attitude of *'après moi, la déluge'*, should be permitted to consider this possibility as well. Although it would be very difficult to find acceptable criteria for selection, the presence of great variability at least provides a favourable starting-point. The realization of such a plan is, however, dependent upon a number of quite radical preconditions: (1) planned procedure with a particular breeding aim carried over several generations; (2) voluntary (or compulsory) renunciation of individual, personal decisions as to whether and with whom one has children (and consequently introduction of artificial insemination); (3) uniform procedure over the entire globe. The third point is particularly important. If one group of people wished to proceed 'as a good example' and to attempt to produce only peaceful and serviceable descendants, they would become easy prey for any neighbour with different inclinations – just like the predator-free and therefore, by nature, peaceable and fearless animals of the Galapagos islands. Such a group of people would be wiped out, or subjected as an uncomplaining slave caste to a society with a retained primaeval drive repertoire. Good intentions would, in either event, go astray.

So the way of planned self-selection is presumably utopian for the present. Without a doubt, humanity could be bettered or exalted in almost every conceivable sense. But it would appear to be impossible for man, in his present state of imperfection, to act as the decisive breeder himself. Consequently, for some time to come, we must attempt to come to terms with ourselves as we now are, whether this be through the principle of world government or through some other means. If this is not successful, then one day somebody will press a particular button operating the most recent and most far-reaching arm of technology. And then, quite abruptly, all life will disappear from the Earth.

14 On freedom   *1961*

The word 'freedom' has a wide range of implications in daily life. If one asks different people what this word calls to mind, the reply might concern political freedom, the solution of internal spiritual compulsions, freedom for actions dictated on ethological or religious grounds, or the independence of voluntary actions from natural law − according to whether the informant is influenced by newspapers, psychoanalytical factors, religious ties or philosophical interests. If the biologist, who is concerned with many different living organisms, is to make some statement about the problem of freedom, it must first be decided whether biology itself − as a theory of life, or as part of natural science − conducts its adherents to a particular aspect of the problem, or at least gives some clues. The first question is the easier to answer, if it is previously recognized that the different applications of the concept of freedom fall into no more than two large categories. The first involves interpretation of freedom as an experience, as a primary property of the ego; it is usually expressed as the concept of free will. The second includes all genuinely or supposedly objectively determinable forms of freedom; that is, the liberation from some compulsion, such as that of following a natural law. The natural scientist, biologist or behavioural research worker is fundamentally unable to comment on the first category, in which freedom occurs as an experience. No process exists to permit us to penetrate the experience, the soul of other organisms and to share the same experience. All attempts to define specific behavioural criteria as proof for (or proof against) experienced free will are entirely arbitrary, lacking any power of conviction. For example, if a hungry cock is confronted with a morsel of food and might take it himself or summon his hens, when he decides upon one of these alternatives after some hesitation it would be quite unfounded to conclude that free will had been operating. Conversely, the oft-repeated claim that animals cannot exhibit any form of free will, since their activities are determined by drives, is just as unacceptable. Who can say that they do not feel themselves to be acting in complete freedom, just like a human being who follows his drive to quench his thirst, to go to sleep or to leap into a river to save a drowning child? Claims for and against experience of freedom by animal organisms permit no conclusions whatsoever about the claimed factual evidence, though they do often permit conclusions about the intellectual make-up of the claimant.

What has been said is also fundamentally valid for conclusions about our fellow men, whose experience is the most accessible to our understanding through extremely vague sympathetic experience. Over two thousand five hundred years ago, the wise man Kuan Tse said: 'See how happy the fish are in the water!' His companion countered: 'How do you know that they are happy; you are not a fish?', and Kuan Tse replied: 'How do you know that I do not know; you are not me.' This brief conversation wonderfully illustrates the inaccessibility of the subjective experience of other organisms; and no knowledge or technology can alter this in any way.

It is therefore a matter of 'freedom' for anybody, including the student of animals, to form his own personal opinion about the experience of animals – and I am divulging no secret when I say that the most successful observers of animals in particular feel themselves constrained to attribute experience to higher animals and to regard as an offshoot of human arrogance the claim that animals are chained by their drives, whilst only man is free.

We must take a closer look at the second category, in which freedom is regarded as a measurable phenomenon, since natural science (biology to a lesser extent than physics) has actually played an active part in opinion formation, and it is still doing so today. I must, however, say 'unfortunately', since much confusion would have been avoided if the scientists and their disciples had rigidly kept to the clear boundaries of their competence.

The Behavioural Physiology of Animals and Man

As physics began to unfold during the last century, there was a long period in which there was a widely-held view that all processes follow natural laws and are physically determined (i.e. in principle predictable). Yet what human beings say and do is, of course, also a process, and this is therefore similarly subject to laws. And – now comes the unjustifiable leap – since human beings usually experience their actions as free, the phenomenon of free will must in reality be an illusion, something which does not actually exist and which has no place in Nature. This line of reasoning is particularly fatal when one attempts to follow it through more rigorously than was usually the case. After all, anybody drawing such conclusions normally excluded from consideration his own person and actions, or at least his own thoughts about the problem of freedom. As a result, various absurd tenets emerged, such as the proposition that a criminal is in fact innocent and should not be punished – as if it were not the case (if physical determinance is taken seriously) that the criminal's acts, the experiences of the jurors, the judge's decision, and everything that I myself think and say, are *all* equally determined. Accordingly, everything stays as it was, and just a few words assume a modified, more undefinable implication – the words 'freedom', 'responsibility' and 'guilt'.

However, at the beginning of this century, the tenet of complete determination was rejected by physicists themselves when they discovered microphysical processes which are evidently not determined within certain limits. Some physicists regard it as quite feasible that the descent of an avalanche and similar events may be set off by fine processes of this kind, so that any concept of fundamental predictability of all processes is no longer tenable. In biology, it is primarily the abrupt, usually very small, modifications of hereditary material (i.e. *mutations*) which are based on similar fine processes. In this case, too, one must conclude that any prediction about future animal or human evolution is subject to certain limits.

Nevertheless, we are brought back to the subject of freedom in that some physicists have hit upon the bright idea that such non-determinable quantum leaps can also occur in the human brain, and that this applies whenever an act of free will is experienced. Accordingly, free will would have its determined place among the natural processes of the brain, and physics would have broken the chains which it had itself imposed on human free will some time ago.

Concepts of this kind have in many instances found a wide circle of adherents, and they have been praised as a liberation, as the decisive step of physics out of base materialism into the sphere of the human intellect.

Unfortunately, all of this represents a house of cards which immediately collapses if we take a closer look at what natural science, or physics, really represents. Physics, or natural science in general, represents a specific occupation of the human intellect – one occupation among many. The aim is to introduce into a wide range of measurable phenomena of the external environment some kind of organization according to certain rules of maximum generality. What cannot be recorded and quantified does not belong in this sphere. Thus, all the qualitative aspects of our experience (such as joy, pain, hope, memory, and all values or judgements) are fundamentally inaccessible to natural science. These human fields exist whether physics is present or not; they were there before physics, and they continue alongside it or above it. This area of primary phenomena also includes the experience of free will and all the consequences of this. Whatever natural science may identify or predict within the human brain can have no influence over this non-physical sphere of experience.

My pain and my will remain as they are whatever the man in the white coat, who is simultaneously studying my brain processes, may provide in the way of scientific conclusions. This is why it was pure phantasy when earlier physicists pronounced the opinion that the determinacy of all processes ruled out free will; and it is similarly pure phantasy with an opposite sign when physicists today maintain that the possibility of non-determinance again permits the existence of my free will. Prohibition and permission from this source represent intellectual trespass which cannot affect the primaeval phenomenon of freedom.

Of course – independently of this – one can ask whether non-determined events can occur in the brain. The answer is virtually certain today. Using mild electrical stimuli transmitted through fine wires painlessly introduced into the brain stem, we can control as we wish the motivation and behaviour of animals. On occasions when operations have had to be performed, such experiments (which are quite harmless) have also been carried out on conscious human beings. It has emerged that one can find stimulation points which will, for example, provide the patient with pleasure, make him eager to marry, recall old memories, and so on. And all of this is experienced not as an imposition from outside, but as spontaneous activity of a free ego.

In view of this, we shall probably have to accustom ourselves to an idea which is doubtless astounding for many people – that the experience of free, spontaneous thought or activity is bound to specific forms of activity in specific brain regions, which themselves may be determined in a completely scientific manner. In the words of Schopenhauer, this would mean: Man can indeed do what he wants; but he cannot will what he wants. But this fact can only reaffirm the absurdity of allowing natural science to prohibit or permit our feelings and wishes, our philosophy of life and our ethical principles. However fascinating science may be, we should realize quite clearly that we should not fall into erroneous worship of the power of something which is only good as long as it remains our slave. At least, that is my conviction – as a scientist.

References

ADRIAN, E. D. (1932) The Mechanism of Nervous Action. Philadelphia.
— (1943): Discharges from Vestibular Receptors in the Cat. *J. Physiol.*, 101, 389–407.
ANTONIUS, O. (1937) Über Herdenbildung und Paarungseigentümlichkeiten der Einhufer. *Z. Tierpsychol.*, 1, 259–89.
— (1939a): Nachtrag. *Z. Tierpsychol.*, 2, 115–17.
— (1939b): Über Symbolhandlungen und Verwandtes bei Säugetieren. *Z. Tierpsychol.*, 3, 263–78.
BABAK, E. (1911) Über die provisorischen Atemmechanismen bei Fischembryonen. *Zbl. Physiol.*, 25, 371–4.
BAEUMER, E. (1955) Lebensart des Haushuhns. *Z. Tierpsychol.*, 12, 387–401.
— (1959) Verhaltensstudie über das Haushuhn – dessen Lebensart, 2. Teil. *Z. Tierpsychol.*, 16, 284–96.
BARCROFT, J. and BARRON, D. H. (1937) Movements in Midfoetal Life in the Sheep Embryo. *J. Physiol.*, 91, 329–51.
BARCROFT, J., BARRON, D. H. and WINDLE, W. F. (1936) Some Observations on Genesis of Somatic Movements in Sheep Embryos. *J. Physiol.*, 87, 189–97.
BARKER, D. (1948) The Innervation of the Muscle-Spindle. *Quart. J. Micr. Sci.*, 89, 143–86.
BERNHARD, C. G. and KOGLUND, C. R. (1942) Slow Positive and Negative Ventral Root Potentials Accompanying Extension and

Flexion Evoked by Medullary Stimulation. *Acta Physiol. Scand.,* 14, Suppl. 47, 7.

BERNHARD, C. G. and THERMAN, P. O. (1947) Alternating Facilitation and Inhibition of the Extensor Muscle Activity in Decerebrate Cats. *Acta Physiol. Scand.,* 14, Suppl. 47, 3.

BESSELER, H. (1949) Zum Problem der Tenorgeige. *Musikal. Gegenwartsfragen,* 1, 1–15.

BETHE, A. (1937a) Experimentelle Erzeugung von Störungen der Erregungsleitung und von Alternans- und Periodenbildungen bei Medusen im Vergleich zu ähnlichen Erscheinungen am Wirbeltierherzen. *Z. vergl. Physiol.,* 24, 613–37.

— (1937b) Rhythmik und Periodik, besonders im Hinblick auf die Bewegungen des Herzens und der Meduse. *Pflüg. Arch.,* 239, 41–73.

— (1940) Die biologischen Rhythmusphänomene als selbständige bzw. erzwungene Kippvorgänge betrachtet. *Pflüg. Arch.,* 244, 1–

— and FISCHER, E. (1931) Plastizität und Zentrenlehre. In: *Handbuch der normalen und pathologischen Physiologie,* Berlin, Bd. 15, 1175–1220.

— and KAST, H. (1922) Synergische und reziproke Innervation antagonistischer Muskeln nach Versuchen am Menschen nebst Beobachtungen über ihre Reaktionszeit. *Pflüg. Arch.,* 194, 77–101.

— and THORNER, H. (1933) Koordinationsstudien an vielbeinigen Tieren (Myriapoden). *Pflüg. Arch.,* 232, 409–31.

BÖHM, H. (1950) Vom lebendigen Rhythmus. *Studium generale,* 4, 28–41.

BOYD, I. A. and ROBERTS, T. D. M. (1953) Proprioceptive Discharges from Stretchreceptors in the Kneejoint of the Cat. *J. Physiol.,* 122, 38–58.

BROCK, L. G., ECCLES, J. C. and RALL, W. (1951) Experimental Investigations on the Afferent Fibres in Muscle Nerves. *Proc. Roy. Soc.,* 138, 453–75.

BROWN, G. (1916) Die Reflexfunktionen des Zentralnervensystems unter besonderer Berücksichtigung der rhythmischen Tätigkeit bein Säuger. *Erg. Physiol.,* 15, 480–780.

BUDDENBROCK, W. von (1937) Physiologie der Sinnesorgane und des Nervensystems. In: *Grundriß der vergleichenden Physiologie,* Bd. I, Berlin.

References

CARLETON, A. (1938) Observations on the Problem of the Proprioceptive Innervation of the Tongue. *J. Anat. Lond.*, **72**, 502–7.

COOPER, S. (1953) Muscle Spindles in the Intrinsic Muscles of the Human Tongue. *J. Physiol.*, **122**, 193–202.

DENNY-BROWN, E., GAYLOR, J. B. and UPRUS, V. (1935) Note on the Nature of the Motor Discharge in Shivering. *Brain*, **58**, 233–7.

DERWORT, A. (1938) Untersuchungen über den Zeitablauf figurierter Bewegungen beim Menschen. *Pflüg. Arch.*, **240**, 661–675.

ECCLES, J. C., FATT, P., LANDGREN, S. and WINSBURY, G. J. (1954) Spinal Cord Potentials Generated by Volleys in the Large Muscle Afferents. *J. Physiol.*, **125**, 590–606.

EDWARDS, C. (1955) Changes in Discharge from a Muscle Spindle Produced by Electrotonus in the Sensory Nerve. *J. Physiol.*, **127**, 636–40.

FISCHER, M. H. (1926) Die Funktion des Vestibularapparates (der Bogengänge und Otolithen) bei Fischen, Amphibien, Reptilien und Vögeln. In: *Handbuch der normalen und pathologischen Physiologie*, Bd. XI, Berlin.

FISCHER, M. H. and WODAK, Z. (1922) Experimentelle Beiträge zu den vestibularen Tonusreaktionen. *Z. Hals-, Nasen-, Ohren-Heilk.*, **3**, 406.

FOERSTER, O. (1936) *Handbuch der Neurologie*. Berlin.

FULTON, J. F. (1943) *Physiology of the Nervous System*. Philadelphia.

FUORTES, M. G. (1954) Direct Current Stimulation of Motoneurones. *J. Physiol.*, **126**, 494–506.

GOLDSTEIN, K. and RIESE, W. (1925) Über den Einfluß sensibler Hautreize auf die sogenannten vestibulären Reaktionsbewegungen. *Klin. Wschr.*, **4**, 1201–4 and 1250–4.

GRANIT, R. (1950a) Autogenic Inhibition. *EEG. Clin. Neurophysiol.*, **2**, 417–24.

— (1950b) Reflex Self-Regulation of Muscle Contraction and Autogenic Inhibition. *J. Neurophysiol.*, **13**, 351–72.

— and KAADA, B. (1952) Influence of Stimulation of Central Nervous Structures on Muscle Spindles in Cat. *Acta Physiol. Scand.*, **27**, 130–60.

— and STRÖM, G. (1951) Autogenic Modulation of Excitability of Single Ventral Horn Cells. *J. Neurophysiol.*, **14**, 113–32.

GRANIT, R., JOB, C. and KAADA, B. (1952) Activation of Muscle Spindles in Pinna Reflex. *Acta Physiol. Scand.*, **27**, 161–8.

GRAY, J. (1950) The Role of Peripheral Sense Organs during Locomotion in the Vertebrates. *Symp. Soc. f. Exp. Biol.*, **4**, 112–26.

— and LISSMANN, H. W. (1946) Further Observations on the Effect of De-afferentiation on the Locomotory Activity of Amphibian Limbs. *J. exper. Biol.*, **23**, 121–32.

— and SAND, A. (1936a) The Locomotory Rhythm of the Dogfish. *J. exper. Biol.*, **13**, 200–9.

— and SAND, A. (1936b) Spinal Reflexes in the Dogfish. *J. exper. Biol.*, **13**, 210–18.

GRIESMANN, B. (1922) Zur kalorischen Erregung des Ohrlabyrinths. *Zbl. Ohrenheilk.*, **19**, 336–7.

GROSS, W. (1938) Flugechsenprobleme. *Natur und Volk*, **68**, 33–43.

HAAS, W. (1940) 2. Reichswettbewerb für Saalflugmodelle. *Luftfahrt und Schule*, **6**, 22–3.

HERTER, K. (1940) Über das Wesen der Vorzugstemperatur bei Echsen und Nagern. *Z. vergl. Physiol.*, **28**, 358–88.

— (1941) Die Vorzugstemperatur bei Landtieren. *Naturwiss.*, **29**, 155–164.

HERTZ, M. (1934) Zur Physiologie des Formen- und Bewegungssehens I. *Z. vergl. Physiol.*, **20**, 430–49.

HESS, W. R. (1954) *Das Zwischenhirn*. 2. Aufl., Basel.

HODES, R. (1953) The Innervation of Skeletal Muscle. *Annual rev. Physiol.*, **15**, 139–64.

HOFFMANN, F. B. (1924) Augenbewegung und relative optische Lokalisation. *Z. Biol.*, **80**, 81–90.

HOLST, E. von (1932) Untersuchungen über die Funktionen des Zentralnervensystems beim Regenwurm. *Zool. Jahrb., Abt. Allg. Zool. u. Physiol.*, **51**, 547–88.

— (1933) Weitere Versuche zum nervösen Mechanismus der Bewegung beim Regenwurm. *Zool. Jahrb., Abt. Allg. Zool. u. Physiol.*, **53**, 67–100.

— (1934a) Über das Laufen der Hundertfüßer. *Zool. Jahrb., Abt. Allg. Zool. u. Physiol.*, **54**, 157–79.

— (1934b) Über die Ordnung und Umordnung der Beinbewegungen bei Hundertfüßern. *Pflüg. Arch.*, **234**, 101–13.

— (1934c) Studien über Reflexe und Rhythmen beim Goldfisch. *Z. vergl. Physiol.*, **20**, 582–99.

— (1935a) Alles oder Nichts, Block, Alternans, Bigemini und verwandte Phänomene als Eigenschaften des Rückenmarks. *Pflüg. Arch.*, **236**, 515–32.

— (1935b) Erregungsbildung und Erregungsleitung im Fischrückenmark. *Pflüg. Arch.*, **235**, 345–59.

— (1935c) Die Gleichgewichtssinne der Fische. *Verh. Dt. Zool. Ges.*, 1935, 108–14.

— (1935d) Die Koordination der Bewegung bei den Arthropoden in Abhängigkeit von zentralen und peripheren Bedingungen. *Biol. Reviews*, **10**, 234.

— (1935e) Über den Lichtrückenreflex bei Fischen. *Pubbl. Staz. Zool. Napoli*, **15**, 143–58.

— (1935f) Über den Prozeß der zentralnervösen Koordination. *Pflüg. Arch.*, **236**, 149–58.

— (1935g) Zentralnervensystem. *Fortschr. d. Zool.*, 1, 364–83.

— (1936a) Vom Dualismus der motorischen und der automatisch-rhythmischen Funktion im Rückenmark und vom Wesen des automatischen Rhythmus. *Pflüg. Arch.*, **237**, 356–78.

— (1936b) Über den 'Magnet-Effekt' als koordinierendes Prinzip im Rückenmark. *Pflüg. Arch.*, **237**, 655–82.

— (1936c) Versuche zur Theorie der relativen Koordination. *Pflüg. Arch.*, **237**, 93–121.

— (1936d) Zentralnervensystem. *Fortschr. d. Zool.*, **2**, 445–68.

— (1937a) Bausteine zu einer vergleichenden Physiologie der lokomotorischen Reflexe bei Fischen II. *Z. vergl. Physiol.*, **24**, 532–62.

— (1937b) Ein neues Instrument für Feinoperationen. *Zool. Anz.*, 119, 328–31.

— (1937c) Vom Wesen der Ordnung im Zentralnervensystem. *Naturwiss.*, **25**, 625–31 and 641–7.

— (1937d) Zentralnervensystem. *Fortschr. d. Zool.*, 3, 343–62.

— (1938a) Neue Versuche zur Deutung der relativen Koordination bei Fischen. *Pflüg. Arch.*, **240**, 1–43.

— (1938b) Über relative Koordination bei Säugern und beim Menschen. *Pflüg. Arch.*, **240**, 44–9.

— (1939a) Entwurf eines Systems der lokomotorischen Periodenbildung bei Fischen. *Z. vergl. Physiol.*, **26**, 481–528 (Habilitationsschrift).

— (1939b) Über die nervöse Funktionsstruktur des rhythmisch tätigen Fischrükkenmarks. *Pflüg. Arch.*, **241**, 569–611.

H O L S T, E. von (1939c) Die relative Koordination als Phänomen und als Methode zentralnervöser Funktionsanalyse. *Erg. Physiol.*, 42, 228–306.

— (1940) Tierflug and Modellflug. Sinn und Bedeutung des Vollschwingen-Flugmodellbaues. *Luftfahrt und Schule*, 3, 24–7.

— (1943a) Über 'künstliche Vögel' als Mittel sum Studium des Vogelflugs. *Journ. Ornith.*, 91, 406–47.

— (1943b) Untersuchungen über Flugbiophysik. I. Messungen zur Aerodynamik kleiner schwingender Flügel. *Biol. Zbl.*, 63, 289–326.

— (1948) Von der Mathematik der nervösen Ordnungsleistungen. *Experientia*, 4, 374–81.

— (1950a) Die Arbeitsweise des Statolithenapparates bei Fischen. *Z. vergl. Physiol.*, 32, 60–120.

— (1950b) Quantitative Messung von Stimmungen im Verhalten der Fische. *Symp. Soc. Exp. Biol.*, 4, 143–72.

— (1950c) Die Tätigkeit des Statolithenapparates im Wirbeltierlabyrinth. *Naturwiss.*, 37, 265–72.

— (1953a) Von der Kunst des Geigenbaues. *Lebendiges Wissen*, Bd. 99 der Sammlung Dieterich. Wiesbaden, 209–15.

— (1953b) Ein neuer Vorschlag zur Lösung des Bratschenproblems. *Instrumentenbau-Z.*, 8, 46–8.

— (1955a) Albrecht Bethe. *Naturwiss.*, 42, 165–7.

— (1955b) Periodisch-rhythmische Vorgänge in der Motorik. *Conf. Soc. Biol. rhythm.*, 5, Stockholm, 7–15.

— (1956) Zentralnervensystem (Das Muskelspindelsystem der Säuger). *Fortschr. d. Zool.*, 10, 381–90.

— (1957a) Aktive Leistungen der menschlichen Gesichtswahrnehmung. *Studium Generale*, 10, 231–43.

— (1957b) Der Saurierflug. *Paläont. Z.*, 31, 15–22.

— (1958) Zentralnervensystem (Die Funktionsstruktur des Zwischenhirns). *Fortschr. d. Zool.*, 11, 245–75.

— and G Ö L D N E R, G. (1953) Eine Vorrichtung zum schnellen Wechsel von Filmszenen (in beliebiger Folge). *Experientia*, 9, 470–2.

— and J E S C H O R E K, W. (1956) Fernreizung freibeweglicher Tiere. *Naturwiss.*, 43, 455.

— and K Ü C H E M A N N, D. (1941) Biologische und aerodynamische Probleme des Tierflugs. *Naturwiss.*, 29, 348–62.

— and K ü h m e, L. (1962) Taktile Täuschungen bei Dickenschätzungen durch Betasten. *Pflüg. Arch.*, **275**, 588–93.

— and L e M a r e, D. W. (1936) Bausteine zu einer vergleichenden Physiologie der lokomotorischen Reflexe bei Fischen I. *Z. vergl. Physiol.*, **23**, 223–36.

— and M i t t e l s t a e d t, H. (1950) Das Reafferenzprinzip. *Naturwiss.*, **37**, 464–76.

— and von S t P a u l, U. (1960) Vom Wirkungsgefüge der Triebe. *Naturwiss.*, **47**, 409–22.

H o l z a p f e l, M. (1939) Analyse des Sperrens und Pickens in der Entwicklung des Stars. *Journ. Ornith.*, **87**, 525–53.

— (1940) Triebbedingte Ruhezustände als Ziel von Appetenzhandlungen. *Naturwiss.*, **28**, 273–80.

H u n t, C. C. and K u f f l e r, S. W. (1951a) The Reflex Activity of Mammalian Small Nerve-fibres. *J. Physiol.*, **115**, 456–69.

— (1951b) Stretch Receptor Discharges during Muscle Contraction. *J. Physiol.*, **113**, 298–315.

— (1952a) The Effect of Stretch Receptors from Muscle on the Discharge of Motoneurons. *J. Physiol.*, **117**, 359–79.

— (1952b) Peripheral Origins of Nervous Activity. *Spring Harbour Symp. Quant. Biol.*, **17**, 113–23.

J u n g, R. (1939) Epilepsie und vasomotorische Reaktionen: Elektroencephalogramm, vegetative Vorgänge und Liquordruck beim kleinen epileptischen Anfall. *Z. Neurol.*, **167**, 601–5.

K a f f k a, G. (1931) Die Bedeutung des Behaviorismus für dir vergleichende Psychologie und Biologie. *Z. Kongo*, Hamburg V, 12–16, IV, 1931, Verh. Dtsch. Ges. Psychol., 213–55.

K a t z, B. (1949) The Efferent Regulation of the Muscle Spindle in the Frog. *J. exper. Biol.*, **26**, 201–17.

— (1950) Depolarisation of Sensory Terminals and the Initiation of Impulse in the Muscle Spindle. *J. Physiol.*, **111**, 261–82.

K a t z, D. (1948) *Die Gestaltpsychologie*. Basel.

K l e m m, O. (1919) *Sinnestäuschungen. Psychologie und experimentelle Pädagogik in Einzeldarstellungen* (Hrsg. R. Schulze), 2 vols. Leipzig.

K o e h l e r, W. (1923) *Die physischen Gestalten in Ruhe und im statischen Zustand*. Braunschweig.

— (1933) *Psychologische Probleme*. Berlin.

K o r n m ü l l e r, A. E. (1932) Eine experimentelle Anästhesie der äußeren Augenmuskeln am Menschen und ihre Auswirkungen. *J. Psychol. u. Neur.*, 41, 354–66.

KORNMÜLLER, A. E. (1947) *Die Elemente der nervösen Tätigkeit.* Stuttgart.

KRAMER, G. (1937) Beobachtungen über Paarungsbiologie und soziales Verhalten der Mauereidechsen. *Z. Morph. Ökol. d. Tiere,* 32, 752–83.

KUFFLER, S. W. (1953) The Two Skeletal Nerve-Muscle Systems in Frog. *Arch. exper. Path. u. Pharm.*, 220, 116–35.

— (1954) Mechanism of Activation and Motor Control of Stretch Receptors in Lobster and Crayfish. *J. Neurophysiol.*, 17, 558–73.

— HUNT, C. and QUILLIAN, J. (1951) Efferent Muscle Spindle Innervation. *J. Neurophysiol.*, 14, 29–54.

— and WILLIAMS, V. (1953) Properties of the Slow-Skeletal Muscle Fibres of the Frog. *J. Physiol.*, 121, 318–40.

KÜSSNER, H. G. (1936) Zusammenfassender Bericht über den instationären Auftrieb von Flügeln. *Luftfahrtforsch.*, 13, 410–24.

— (1940) Allgemeine Tragflächentheorie. *Luftfahrtforsch.*, 17, 370–8.

LE MARE, D. W. (1936) Reflex and Rhythmical Movements in the Dogfish. *J. exper. Biol.*, 13, 429–42.

LILIENTHAL, O. (1939) *Der Vogelflug als Grundlage der Fliegekunst.* 3rd edn., München–Berlin.

LIPPISCH, A. (1925) Theoretische Grundlagen des Schwingenfluges. *Flugsport*, 17, 246–53.

— (1935) Gedanken zum Schwingenflug. *Luftwelt*, 2, 106–7 (Heft 3).

— (1936) Schwingenflug. *Der Segelflieger*, 11, 11–12 (Heft 9), 10–11 (Heft 10), 9–10 (Heft 11), 10–11 (Heft 12).

LISSMANN, H. W. (1946) The Neurological Basis of the Locomotory Rhythm in the Spinal Dogfish (*Scyllum canicula, Acanthias vulgaris*). *J. exper. Biol.*, 23, 162–76.

LLOYD, D. P. C. (1953) Influence of Asphyxia upon the Responses of Spinal Motoneurons. *J. Gen. Physiol.*, 36, 673–702.

LORENZ, K. (1933) Beobachtetes über das Fliegen der Vögel und über die Beziehungen der Flügel- und Steuerform zur Art des Fluges. *Journ. Ornith.*, 81, 107–236.

— (1937a) Über den Begriff der Instinkthandlung. *Fol. Biotheoret.*, 2, 17–50.

— (1937b) Über die Bildung des Instinktbegriffes. *Naturwiss.*, 25, 289–300, 307–18, 324–31.

— (1937c) Biologische Fragestellung in der Tierpsychologie. *Z. Tierpsychol.*, **1**, 24–32.

— (1939a) Die Paarbildung beim Kolkraben. *Z. Tierpsychol.*, **3**, 278–292.

— (1939b) Vergleichende Verhaltensforschung. *Verh. Dt. Zool. Ges.*, Suppl. 12, 69–102.

— (1940) Durch Domestikation verursachte Störungen arteigenen Verhaltens. *Z. angew. Psychol. u. Charakterk.*, **59**, 1–81.

— and TINBERGEN, N. (1939) Taxis und Instinkthandlung in der Eirollbewegung der Graugans. *Z. Tierpsychol.*, **2**, 1–29.

LOWENSTEIN, O. (1950) Labyrinth and Equilibrium. *Symp. Soc. f. Exp. Biol.*, **4**, 60–82.

— and ROBERTS, T. D. M. (1949) The Equilibrium Function of the Otolith Organs of the Thornback Ray (*Raja clavata*). *J. Physiol.*, **110**, 392–415.

McCOUCH, G. P., DEERING, D. and LING, T. H. (1951) Location of Receptors of Tonic Neck Reflexes. *J. Neurophysiol.*, **14**, 191–5.

MAGNAN, A. (1931) *Le vol des oiseaux et le vol des avions.* Paris.

MAGNUS, R. (1924) *Körperstellung.* Berlin.

MASCOW, H. (1940) *Vom Vogelflug zum Menschenflug.* Neiße.

MATTHAEI, R. (1929) Das Gestaltproblem. *Erg. Physiol.*, **29**, 1–82.

MATTHEWS, B. H. G. (1933) Nerve Endings in Mammalian Muscle. *J. Physiol.*, **78**, 1–53.

MITTELSTAEDT, H. (1949) Telotaxis und Optomotorik von Eristalis bei Augeninversion. *Naturwiss.*, **36**, 90–1.

— (1950) Physiologie des Gleichgewichtssinnes bei fliegenden Libellen. *Z. vergl. Physiol.*, **32**, 422–63.

— (1954a) Regelung in der Biologie. *Regelungstechnik*, **2**, 177–81.

— (1954b) Regelung und Steuerung bei der Orientierung von Lebewesen. *Regelungstechnik*, **2**, 226–32.

MOUNTCASTLE, V. B., COVIAN, M. R. and HARRISON, C. R. (1952) The Central Representation of some Forms of Deep Sensibility. *Res. Publ. Ass. nerv. ment. Dis.*, **30**, 339–70.

MYGIND, S. H. (1948) Static Function of the Labyrinth. Attempts at a Synthesis. *Acta oro-laryng.*, Suppl. 70, Kopenhagen.

OLEFIRENKO, P. P. (1937) Über periodische, mit dem Atemrhythmus synchrone Extensionsbewegungen der Extremitäten bei decerebrierten Hunden. *Fizill. Z.*, (Russ.) **23**, 24–33.

PEIPER, A. (1938) Das Zusammenspiel des Saugzentrums mit dem

Atemzentrum beim menschlichen Säugling. *Pflüg. Arch.*, **240**, 312–24.

PELKWIJK, J. J. and TINBERGEN, N. (1937) Eine reizbiologische Analyse einiger Verhaltensweisen von *Gasterosteus aculeatus*. *Z. Tierpsychol.*, **1**, 193–200.

PETERS, H. (1937) Experimentelle Untersuchungen über die Brutpflege von *Haplochromis multicolor*, einem maulbrütenden Knochenfisch. *Z. Tierpsychol.*, **1**, 201–18.

RUFFINI, A. (1893a) Sulla terminazione nervosa nei fusi musculorie sul loro significato fisiologico. *Atti Accad. naz. Lincei* (5) Rendic. Vol. **1**, 2. Sem. Fasc. 1, 31–8.

— (1893b) Di una particolare reticella nervosa e di alcuni corpuscoli di Pacini che si trovano in connesione cogli organi musculotendinei del gatto. *Atti Accad. naz. Lincei* (5) Rendic. Vol. **1**, 1. Sem. Fasc. 12, 442–6.

— (1893c) Sur un réticule nerveux spécial, et sur quelques corpuscules de Pacini qui se trouvent en connexion avec les organs musculo-tendineux du chat. *Arch. Ital. Bid. T.*, **18**, Fasc. 1, 101–114.

SCHMIDT, R. (1939) *Flug und Flieger im Pflanzen- und Tierreich*. Berlin.

SCHOEN, L. (1949) Quantitative Untersuchungen über die zentrale Kompensation. *Z. vergl. Physiol.*, **32**, 121–50.

SCHÖNE, H. (1950) Die Augen als Gleichgewichtsorgane bei Wasserkäferlarven. *Naturwiss.*, **37**, 235–6.

SEITZ, A. (1940) Die Paarbildung bei einigen Cichliden. I. Die Paarbildung bei *Astatotilapia strigigena* Pfeffer. *Z. Tierpsychol.*, **4**, 40–84.

SHERRINGTON, C. S. (1892) Notes on the Arrangement of some Motor Fibres in the Lumbosacral Plexus. *J. Physiol.*, **13**, 621–772.

SPIEGEL, E. A. and SATO, G. (1927) Experimentalstudien am Nervensystem. *Pflüg. Arch.*, **215**, 106–32.

SPRAGUE, J. M., SCHREINER, L. H., PINDSLEY, D. B. and MAGOUN, W. (1948) Reticulo-spinal Influences on Stretch Reflexes. *J. Neurophysiol.*, **11**, 501–7.

STOLPE, M. and ZIMMER, K. (1939) Der Schwirrflug des Kolibri im Zeitlupenfilm. *Journ. Ornith.*, **87**, 136–55.

STOPFORD, J. S. B. (1921) The Nerve Supply of the Interphalangeal and Metarcarpo-Phalangeal Joints. *J. Anat. London*, **56**, 1–11.

STRESEMANN, E. (1931) Aves. In: *Handbuch der Zoologie*, 572–880.

TAUSCH, R. (1954) Optische Täuschungen als artifizielle Effekte der Gestaltungsprozesse von Größen und Formenkonstanz in der natürlichen Raumwahrnehmung. *Psychol. Forschung,* 24, 299–348.

TINBERGEN, N. (1940) Die Übersprungbewegung. *Z. Tierpsychol.,* 4, 1–40.

— and KUENEN, D. J. (1939) Über die auslösenden und richtunggebenden Reizsituationen der Sperrbewegung von jungen Drosseln. *Z. Tierpsychol.,* 3, 37–60.

TÖNNIES, J. F. (1949) Die Erregungssteuerung im Zentralnervensystem. *Arch. Psychiatr. u. Z. Neur.,* 182, 478–535.

TRENDELENBURG, W. (1906) Über die Bewegung der Vögel nach Durchschneidung hinterer Rückenmarkswurzeln. Ein Beitrag zur Physiologie des Zentralnervensystems der Vögel (Nach Untersuchungen an *Columba domestica*). *Arch. Anat. u. Physiol.,* 1906, 1–126.

— (1907) Zur Kenntnis der Physiologie der Skelettmuskulatur. *Arch. Anat. u. Physiol.,* 1907, 499–506.

— (1943) *Der Gesichtssinn.* Berlin.

VOLKELT, H. (1937) Tierpsychologie als genetische Ganzheitspsychologie. *Z. Tierpsychol.,* 1, 49–65.

WEISS, P. (1936) Selectivity Controlling the Central-Peripheral Relations in the Nervous System. *Biol. Rev. Cambridge philos. Soc.,* 11, 494–531.

— (1941) Selfdifferentiation of the Basic Patterns of Coordination. *Comp. Psychol. Monogr.,* 17, 1–96.

WERNER, C. L. F. (1940) *Das Labyrinth. Bau, Funktionen und Krankheiten des Innenohres vom Standpunkt einer experimentellen und vergleichenden Pathologie.* Leipzig.

WINKLER, H. (1940) *Modellflug,* 6, 2.

Index

Note : All references to illustrations are printed in italic.

331

Index

automatic-rhythmic element, 90, 104
automatism, 11, 13f., 29, 87, 92, 94, 100, 105, 110, 113. *See also* endogenous activity; spontaneous activity
averaging (CNS), *243*

Babak, E., 115
balancing movement, *see* equilibrating movement
Barany, 170
barbiturate, 310
Barcroft, J., 115
Barker, D., 175, *176*
bat, 286–7
beak-cleaning, *247*
bee, 272, 288
behaviour pattern, 300. *See also* instinctive behaviour pattern; innate behaviour
behavioural inventory, 232
behavioural science, 218, 277, 314
behavioural sequence, 240, 246f., 256
behaviourism, 132
Berger, H., 129
Bernhard, C. G., 147
Bethe, A., 5, 35, 103, 112, 157, 171
bigemini, 112
biological clock, 280–1
bird, 149, 168, 275, 289
bird flight, 35, 103
birth control, 311–12
blackbird, 274
blinking, 247, 249
Böhm, H., 150
Boyd, I. A., 183
brain, 41, 106, 117, 126, 221, *223*, 233, 279–82, 288, 292, 296, 298, 302–3, 309, 310, 316
  disorders of, 36, 296
  electronic, 294
  extirpation, 6–8
  See also central nervous system; midbrain
brainstem, 179, *223*, 225, *239*, 257, 302–3, 316–17
brightness constancy, 193–4, 196
Brock, L. G., 179
brooding, 232, 237, *243*, 244, 252
Brown, G., 5, 83
bullfinch, 274
Burckhardt, J., 307
Burmese jungle fowl, 276

cackling, *227*, 235, 240–1, *242*, 245–6, 252, 276
call, *see* vocalization
calling to food, *224*, 301
calling to nest, 238
cancellation (CNS), *243*, 244
cannibalism, 306
carbon dioxide supply, 29, 43, 115
Carleton, A., 179
cat, 175, 177, 179, 183, 221
causality, 296
centipedes, 6, 35
central activity state, 128
central adaptation, 232, *233*, *235*, 239–40, 252, 254, 255–6
central coordination, 31, 33f., 104, 108, 112, 128, 140, 189, 288
central elements, 84, 103–4
'central feedback', 151
central nervous processes, 64, 73, 83, 84, 86, 90–1, 102, 127, 131, 134, 191, 214, 279, 282, 317
central nervous system, 20, 27–8, 29, 31, 33–4, 45, 69, 73, 120–3, 126, 127, 129, 131–2, 139–42, 160, 162, 164, 165, 168–9, 171–3, 174f., 220f.
central rhythm, 47, 49, 60–1, 66, 75, 79, 81f., 82, 84, 86. *See also* fin rhythms
central superimposition, *see* superimposition
'centres', 90–1, 145, *146*, 148, 150, *153*, *166*, 172, 174, 199, 221, 284, 310. *See also* localization
centrifugal field, 144, *145*, 160, 182, 195
cerebral cortex, 129, 179, 286
chain-reflex, 3, 31, 82, 99
cheek-scratching, 247
Cheyne-Stokes periodicity, 123
  in respiration, 115
chick, 289
chicken, 220f., *239*, *247*, *248*, *250*, *253*, 300, 302
cock, *227*, 238, *239*, *242*, 247, *250*, 301, 312, 315
  hen, 230, *253*, 275, 315
clucking, *224*, 226, *229*, *233*, *234*, *238*, *240*, 241, *242*, 245–6, *248*, 251–2, *254*, 255, 257
CNS, *see* central nervous system
coactive position (rhythms), 63–6, 72, 80, 94, 106–7, 123–5, 134–5
cock, *see* chicken

Index

334

Index

Index

Index

spontaneously produced perception, 196. *See also* illusion

Sprague, J. M., 179

stable equilibrium in nervous coordination, 62

standing up, 229, 230, 244, *245, 248, 249,* 252, 257, 302

starling, 274, 276

statolith, 143, 144, 145, *146,* 147, 168, 195, 196, 283, 284

removal of, 144, 148, *169*

See also labyrinth system

Steiner, G., 306

'steered reflex', 143, 144

stick-insect, 6

stimulation field (CNS), 235, 236, 237, 239, 243, 246, 247, 248, 249, 252, 255, 256, 302

stimulation of the nervous system (artificial), 174, 177, 179, 195, 221, *223,* 224, *229, 250,* 284, 302, 309, 317

stimuli, 34, 42, 43, 47–9, 50, *54,* 64, 83, 84, 89, 96, 102, 117, 128, *130,* 132, 139, 140, 164, 170, 171, 175, 179, 182, 190, 192, 218, 243, 275, 276, 300, 301

inhibitory, 37, 48

supplementary, 112, *113*

See also inhibition; peripheral stimuli; pain stimuli; reflex action

stimulus field, *see* stimulation field

stimulus intensity, 87

stimulus strength (electrical), 231

stimulus time (electrode), 227, 229, 231, 233, *241*

stimulus voltage, 234, 242

Stopford, J. S. B., 183

stretch-receptor, 178. *See also* tension receptor

strychnine, 288

effects on fin rhythms, 47, *48,* 94, 96, 103, 114, 123

subconscious process, *see* non-conscious process

submission, 251, 277, 291

subthreshold stimulation, 236

suckling of human infant, 123f.

summation of nervous processes, 51, 76, 80, 151, 170, 187, 234, 236. *See also* quantitative interaction

superimposition, 11–13, 15, 17, 20, 22– 24, 27, *28,* 30, 74f., *75,* 77, *78, 81,* 84, 85, 89, 94, 105, 108, 110, 111, 117, 118, 122, 124, 130, 132, 133, 151

interaction with magnet effect, 80, 105, 108

superimposition hypothesis, 75–7, 119

superposition (CNS), 240, *243*

suppression of action, 67, 73, 82, 89, 104, 129, 244, 308, 309. *See also* inhibition

swallowing, 251

swimming

fish, *145*

man, 122

switching mechanism, 143

synapse, 179

synergy, 102, 103, 104, 105, 179, *180*

tabes dorsalis, 163–4

tape-recorded calls, 246

Tausch, R., 212

technology, 294, 307, 312, 315

temperature

effect on fin rhythms, *46,* 50, 64, 96

effect on tactile sensation, *263*

temporal distance (fin rays), 97, 98, 100

temporal distance table, 97, *98,* 99

temporal lobe, 282

tench, 5, 82

tendon, 160, 162, 178

receptor, 160, 177, 178, 179, 181, 182

spindle, *161,* 261

tension, 160, 178, 179, 183

receptor, *161,* 162, 163

See also tendon

tensor tympani, 176

territory, 311

tetanic contraction, 158

Therman, P. O., 147

thirst, 237, 253, 257, 300, 302

Thorner, H., 35

threat, 238, 252, *253,* 255, 311

threshold to stimulation, 162, 178, 224, 225, 228, 233, 234, 236, 237, 240, *245,* 246, 247, 255, 256

threshold voltage, 225, 226, 229, 231

'time and speed tables', 24, 57f., 74, *75,* 81, 99

definition of, 58

Tinbergen, N., 129, 140

Science

Date Due

DEC 2 1994			